Monsoon
Meteorology

Mature cumulonimbus northeast of a U.S. Weather Bureau Research Flight Facility aircraft. Aircraft flight level: 5550 m; position: 15°28′N, 87°55′E; time: 0650 GMT, 1 June 1963. The cloud was part of a 200-km-long NW–SE line of cumulonimbus lying 200 km west of the nimbostratus cloud mass of a subtropical cyclone centered near 11N, 95E (see Fig. 3.10).

Monsoon Meteorology

C. S. Ramage

UNIVERSITY OF HAWAII
HONOLULU, HAWAII

 1971

ACADEMIC PRESS · New York and London

ACADEMIC PRESS, INC.
111 Fifth Avenue, New York, New York 10003

United Kingdom Edition published by
ACADEMIC PRESS, INC. (LONDON) LTD.
Berkeley Square House, London W1X 6BA

LIBRARY OF CONGRESS CATALOG CARD NUMBER: 73-127697

PRINTED IN THE UNITED STATES OF AMERICA

This is Volume 15 in
INTERNATIONAL GEOPHYSICS SERIES
A series of monographs
Edited by J. VAN MIEGHEM, *Royal Belgian Meteorological Institute*

A complete list of the books in this series appears at the end of this volume.

I have been to Bombay so many times, but I had never seen the coming of the monsoon there. I had been told and I had read that this coming of the first rains was an event in Bombay; they came with pomp and circumstance and overwhelmed the city with their lavish gift. It rains hard in most parts of India during the monsoon and we all know this. But it was different in Bombay, they said; there was a ferocity in this sudden first meeting of the rain-laden clouds with land. The dry land was lashed by the pouring torrents and converted into a temporary sea. Bombay was not static then; it became elemental, dynamic, changing.

So I looked forward to the coming of the monsoon and I became a watcher of the skies, waiting to spot the heralds that preceded the attack. A few showers came. Oh, that was nothing, I was told; the monsoon has yet to come. Heavier rains followed, but I ignored them and waited for some extraordinary happening. While I waited I learnt from various people that the monsoon had definitely come and established itself. Where was the pomp and circumstance and the glory of the attack, and the combat between cloud and land, and the surging and lashing sea? Like a thief in the night the monsoon had come to Bombay, as well it might have done in Allahabad or elsewhere. Another illusion gone.

Jawaharlal Nehru, *The Monsoon comes to Bombay*, 1939.

Contents

Preface

The facts of monsoon climate have been presented in a number of texts and atlases. Geographers and climatologists have written books on the monsoons. While climate and statistics were treated, these chroniclers seldom accounted for normals, standard deviations, and variabilities in terms of the day-to-day weather of which climate is made. Most emphasis was placed on the solstices; conditions at the height of summer and, more rarely, at the height of winter were described. The radical transformations between solstices were not discussed.

This book is constructed around synoptic systems common in the monsoon area. Differing sequences and frequencies of these systems are related to both regional and interannual variations in the character of the monsoons. Changes in the mix of synoptic systems through the transition seasons must be comprehended too, as essential parts of the dramatic annual cycle which is the monsoon hallmark. Using synoptic models as building blocks, a monsoon area structure of linked diversity becomes possible.

This book is directed specifically to upper division undergraduate or graduate students of meteorology or climatology who have studied theoretical meteorology for at least a year, and to professional meteorologists working or expecting to work in the monsoon area.

It may also prove useful to the growing number of meteorologists studying daily changes of the atmosphere on a global scale.

The present monograph is neither a compendium of solutions nor a handbook of tested procedures; I discourage expectation of easy answers; nevertheless, promising monsoon area research remains to be done on every scale from simple local studies to circulation simulations on giant computers.

The amount of research that has been done in monsoon meteorology, though widely scattered, is impressive, but unevenness of quality and quantity has made writing a smooth-flowing book difficult. I have tried to compensate for my own uneven experience, largely confined to Asia and the western Pacific, by emphasizing Africa. I am afraid imbalance remains, however, because of the extensive research in India, China, and Japan.

Monsoon research is provincial. Investigators have seldom been aware that in other monsoon regions similar problems may have been under study or even solved. Inefficient scientific communication partly accounts for this.

The only widely distributed journals are published in middle-latitudes. Regional journals or research reports often are well distributed beyond the monsoon area but poorly distributed within it. I hope that extensive cross-referencing within this text, together with the reference list, will demonstrate to monsoon meteorologists that help in problem-solving is as likely to come from within the area as from beyond it.

Poor communication has also encouraged invention of local terminologies which have quite successfully disguised the fact that similar phenomena occur throughout the monsoon area. Conversely, single terms ("disturbance" for example) have been assigned quite different regional meanings. From this welter I have selected a terminology which I hope unambiguously identifies features without falsely implying physical understanding.

There is never an ideal time to write a book such as this. Sirens singing of tomorrow's wondrous satellite probes and fantastically accurate numerical models celebrate the rewards of delay. But before the promised marvels envelope us, an individual has one last chance to comprehend and try to describe the generality of monsoon meteorology. The job is tiresome, but if indeed a new era is dawning, stock-taking could be useful.

Acknowledgments

Over the past fourteen years the National Science Foundation and the Air Force Cambridge Research Laboratories have generously supported my colleagues and me in our studies of monsoon problems. The Navy Weather Research Facility has aided this book's rapid preparation which has been benignly regarded by the University of Hawaii.

I appreciate the frank, constructive comments proffered by my colleagues in Honolulu, India, Hong Kong, and on the U. S. mainland, having been particularly helped by Gordon Bell, Robin Brody, Edward Carlstead, Wan-cheng Chiu, Takio Murakami, Chester Newton, James Sadler, and Ronald Taylor.

Without the patient efficiency of Mrs. Michie Hamao who typed the drafts and manuscript, Mr. Louis Oda who drafted the diagrams, and Mrs. Ethel McAfee who edited the manuscript, the book would have remained a bundle of scribbled sheets; without my wife's constant cheery encouragement I should have long since backslid to the easy life of writing papers.

Monsoon
Meteorology

> It would be completely wrong to . . . not use the expression
> monsoon at all,—in the voluminous Compendium of Meteor-
> ology, "monsoon" does not even appear in the index, much
> less is a contribution or paragraph devoted to it!— . . .; but also
> the other extreme, represented by the Russian researchers
> Alissov and Khromov, who apply the label "monsoon" to any
> annual change of wind direction on earth, even over the free
> ocean and therefore completely independent of surface con-
> ditions, seems not justified to the author.
>
> Joachim Blüthgen, *Allgemeine Klimageographie,* 1966.

1. Definition of the Monsoons and Their Extent

Just as every viewer has his personal rainbow, so every meteorologist seems to possess a personal and singular understanding of what is meant by "monsoon." All agree that the Indian monsoon is the *fons et origo*, and that it comprises two distinct seasonal circulations—a winter outflow from a cold continental anticylone and a summer inflow into a continental heat low. But what other regions are monsoonal?

For more than a century, meteorologists and geographers have been variously answering this question. Their efforts, as Blüthgen laments, have not led to a consensus. Perforce, then, before embarking on any detailed discussion of monsoons, I need to provide yet another definition, and I hope a rational one to justify writing this book and to delimit its scope. To be fair, I must point out that my definition, although objectively supported, is not entirely objective in origin. It derives in part from thirteen years' residence in the monsoon area.

The word "monsoon" has an ancient and debatable etymology which many authorities have already tried to trace. All agree, however, that the central meaning incorporates seasonality—surface winds flowing persistently from one quarter in summer and just as persistently from a different quarter in winter. Implied is a corresponding change in the direction of the surface-pressure gradient and in prevailing weather.

1.1 Monsoon Index

In order to objectively delimit the monsoons, Hann (1908) defined a "monsoon index," referred to an eight-point surface wind rose. The index

I_H = maximum difference in frequency between midwinter and midsummer (midwinter higher) + maximum difference in frequency between midwinter and midsummer (midsummer higher).*

Schick (1953) proposed a slightly different index:

$$I_S = (F_{Jan} - F_{July}) + (F'_{July} - F'_{Jan})$$

where F and F' are percent frequencies and refer to direction of the prevailing winds for January and July, respectively. The maximum possible value for both I_H and $I_S = 200$; I_S will always be equal to or less than I_H, and the differences could be considerable when the index tends to be low (Table 1.1). Unfortunately neither index is weighted by the *amount* of direction change.

Kao *et al.* (1962) suggested yet another index which removes this difficulty:

Table 1.1
Comparison of Hann, Schick, Kao *et al.*, and Khromov Monsoon Indices[a]

	Frequency (%)							
	N	NE	E	SE	S	SW	W	NW
Bahrein (26°16′N, 50°37′E)[b]								
January	26	3	3	4	17	12	6	28
July	62	4			1	1	1	26
Differences	−36	−1	+3	+4	+16	+11	+5	+2
Chittagong (22°21′N, 91°50′E)[c]								
January	25	4	1		5	6	9	35
July	1	2	2	23	59	10	1	1
Differences	+24	+2	−1	−23	−54	−4	+8	+34

[a] From Hann (1908), Schick (1953), Kao *et al.* (1962), and Khromov (1957).
[b] $I_H = I_K = 36 + 16 = 52$; $I_S = 36 + 2 = 38$; I_{Kh} = nonmonsoonal.
[c] $I_H = I_S = 54 + 34 = 88$; $I_K = 54 + 24 = 78$; $I_{Kh} = (35 + 59)/2 = 47\%$.

* See Table 1.1.

I_K = the sum of the maximum values of the differences between the frequency of occurrence of winds from the same direction and from the opposite direction in January and July.

$I_K = I_H$ for Bahrein but $I_K < I_H$ for Chittagong (Table 1.1). I question whether this "improvement" justifies recomputing monsoon indices, especially over land where orographic channeling so often renders surface winds unrepresentative.

In any case, Khromov (1957) provided a somewhat different, but satisfactory, solution. He first outlined those regions where the prevailing wind direction shifts at least 120 degrees between January and July. Then within these regions, he analyzed the average of the frequencies of the prevailing wind directions in each of the two months:

$$I_{Kh} = (F_{Jan} + F'_{July})/2$$

According to Khromov, Bahrein is nonmonsoonal. In Fig. 1.1, Khromov's analysis is modified slightly through incorporation of recent data.

Khromov assigned to those regions outside the 40% isopleth a monsoon "tendency" and described those within the 40% isopleth as monsoonal. By these criteria, most of southern, northern, and eastern Asia is monsoonal. The 40% isopleth encloses isolated areas over the northeastern, southeastern, and southwestern Pacific affected by the normal annual march of the subtropical ridge. A similar shift occurs in the near-equatorial troughs over the

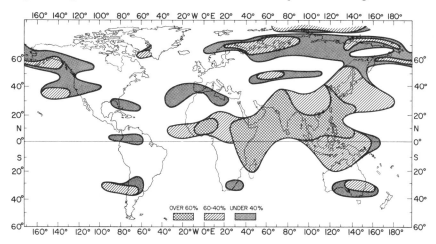

Fig. 1.1. Geographical extent of the monsoons according to Khromov (1957). Average frequency of predominant surface wind directions shown in three categories. Unshaded areas are nonmonsoonal.

central Indian Ocean and over the eastern Pacific west of Mexico. Since in all these areas, mean resultant winds are light and often variable, I consider them to be nonmonsoonal (Fig. 1.2). By adding a *wind strength* criterion to Khromov's direction-based monsoon index I have given some weight to the annual *vector* variation in the circulation. The Americas and Europe are nonmonsoonal.

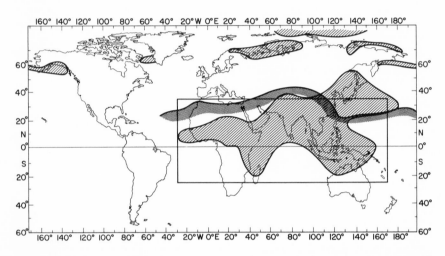

Fig. 1.2. Final delineation of the monsoon region. Hatched areas are "monsoonal" according to Khromov (1957). Heavy line marks northern limit of the region within the Northern Hemisphere with low frequencies of surface cyclone–anticyclone alternations in summer and winter (Klein, 1957). Rectangle encloses the monsoon region.

1.2   Circulation Persistence

The question of monsoons in higher latitudes has been vigorously debated by many meteorologists (cf. Alisov, 1954; Shapaev, 1960; Blüthgen, 1966; Kurashima, 1968).

This brings me to the last part of an adequate monsoon definition: only those subregions, *where surface cyclones and anticyclones alternate infrequently in summer and winter*, are monsoonal. In other words, the seasonal wind shifts should not reflect averaging of a shift in the tracks of moving circulations, but rather the replacement of one *persistent* circulation system by a reverse and equally persistent circulation system. In the following paragraphs I separate middle latitudes from the Arctic whence definitive data have only recently been obtained.

1.2.1   MIDDLE LATITUDES

Ward (1925) observed that, " The weaker general pressure controls and the frequent cyclonic interruptions of . . . seasonal winds in the United States prevent any such marked development as that attained by the monsoons of India." In other words, the United States is widely reckoned not to be monsoonal, although the climate in the west, and south, possesses some monsoonal characteristics (Fig. 1.1).

Klein (1957) reproduced monthly charts of the Northern Hemisphere showing numbers of individual surface cyclone and anticyclone centers recorded over 20 yr in 5° latitude–longitude rectangles. For Texas the sum of January and July numbers ranges upward from 40. The heavy line on Fig. 1.2 marks the northern limit of those regions with totals ranging downward from 40, thereby satisfying the second part of my monsoon definition.

1.2.2   SOUTHERN HEMISPHERE

No such survey as Klein's has been made of the Southern Hemisphere. Nevertheless, Khromov's boundaries are probably correct. In the *southwest*, Frolow (1960) doubted whether Madagascar is monsoonal. In the *southeast*, Taylor (1932) considered northern Australia to be as monsoonal as India; Braak (1921–1929) asserted that, " The influence of Asia and Australia renders [Indonesia] the most typical monsoon region of the world "; but Brookfield and Hart (1966) later classified the southwestern Pacific as nonmonsoonal.

1.2.3   ARCTIC

Is the Arctic monsoonal? Shapaev (1960) claimed that at least the Russian Arctic is, and since the cold pole lies over the Arctic Ocean in summer and over Siberia in winter, there once appeared to be superficial justification for the claim. Klein's data were inadequate to settle the question, but before long two papers appeared utilizing IGY, ice-island, and aircraft reconnaissance observations to delineate the synoptic climatology of the region. Both Keegan (1958) and Reed and Kunkel (1960) reported considerable synoptic activity throughout the year, at least the equal of that in the middle latitudes. The authors clinched the case against an Arctic monsoon with their charts of mean sea-level pressure. In both summer and winter the Arctic Ocean is the seat of relative low pressure.

1.3   Monsoon Definition

I define the monsoon area as encompassing regions with January and July surface circulations in which:

1. the prevailing wind direction shifts by at least 120° between January and July,
2. the average frequency of prevailing wind directions in January and July exceeds 40%
3. the mean resultant winds in at least one of the months exceed 3 m sec^{-1}, and
4. fewer than one cyclone–anticyclone alternation occurs every two years in either month in a 5° latitude–longitude rectangle.

The only region satisfying all parts of my monsoon definition is outlined in Fig. 1.2. Using the South Asian mountains as a natural northern boundary and squaring off, I can enclose the monsoons between 35N and 25S and between 30W and 170E. The subject of this book is now defined and its scope delimited.

1.4   Weather and Climate

As the discussion develops the reader will realize that a paucity of surface disturbances by no means signifies unchanging weather and that, in the height of the monsoons, middle-troposphere synoptic scale systems frequently develop and decay and, less frequently, travel considerable distances. It is as if the intense, but shallow, cold highs of winter and the warm lows of summer can inhibit development of synoptic systems in only the lowest layers of the atmosphere.

I have deliberately avoided mentioning weather as a monsoon criterion, although many writers (c.f. Conrad, 1936) have attempted to graft onto their circulation criteria a "wet summer/dry winter" requirement. The resulting confusion underscores the fact that no simple correlation exists between the direction of the surface pressure gradient and rainfall. Summer continental heat lows are no less arid than winter continental anticyclones.

1.5   Subsequent Chapters

In the next chapter I describe the regional climatology of the surface monsoon circulations, attempting to account for their extent and intensity.

Chapter 3 describes the peculiar synoptic components of the monsoons, which, by their comings and goings, ensure that day-to-day weather during the height of a monsoon seldom exhibits seasonal sameness.

Chapter 4 deals with mesoscale systems, diurnal variations, and differing rain characteristics, which sometimes are parts of and at other times modify the larger synoptic components and their accompanying weather.

In Chapter 5 the preceding three chapters are integrated into a synoptic climatology of the whole year, into which the summer and winter monsoons are fitted and linked through the transition seasons.

Synoptic analysis and short-period forecasting based on analysis or on statistics are the subjects of Chapter 6.

In Chapter 7, I present examples of the role of the monsoons in the atmospheric circulation, and of major monsoon anomalies leading to floods and droughts. The difficulties of long-range monsoon forecasting are treated briefly.

The final summary chapter emphasizes unifying concepts and identifies problems for which research could provide solutions.

Meteorological variables—unspecified with respect to level—refer to the surface layer. In Chapter 3 a few diagrams, reproduced from other works, depict wind speeds in knots (divide by 2 to obtain the approximate meters per second equivalents). Otherwise, the cgs system is used throughout except when heights are expressed, in accordance with meteorological convention, in millibars. In the tropics, approximate pressure-height/kilometer equivalents are as follows:

Pressure height (mb)	1000	850	700	500	300	200	150	100
Height (km)	0	1.5	3.1	5.8	9.6	12.4	14.1	16.5

To ensure continuity in viewing, Plates I–IV are included at the end of the book. Each plate, representing a mid-season month for the whole monsoon area, contains eight charts presenting a condensed three-dimensional view of average circulation and weather.

From the Indian Archipelago across the east coast of Africa there are very different winds and weather at opposite seasons of the year.

Robert Fitzroy, *The Weather Book*, 1863.

2. Regional Climatology of the Surface Monsoon Circulations

The monsoons blow in response to the seasonal change that occurs in the difference in pressure—resulting from the difference in temperature—between land and sea. Where continents border oceans, large temperature differences and hence large differences in pressure might be expected. However, the shapes of the continents and their topographies, as well as variations in sea-surface temperatures, all interact to produce considerable regional and temporal variability in the monsoons. In this chapter I will first discuss large-scale coastal influences, then identify the main monsoon regimes and attempt to account for their intensities and extents, using January and July long-term means of sea-level pressure (Figs. 2.1 and 2.3), resultant surface winds (Plates I and III), surface temperatures (Figs. 2.2 and 2.4), range of ocean-surface heat balance (Fig. 2.5), annual variation of rainfall (Fig. 2.6), and variability of rainfall (Fig. 2.7). Finally, to introduce subsequent chapters, I will delineate the period and scale ranges of monsoon variations.

2.1 Data

Excellent marine climatological atlases (McDonald, 1938; Meteorological Office, 1947, 1949; van Duijnen Montijn, 1952; U.S. Weather Bureau, 1955–1959; Deutsches Hydrographisches Institut, 1960) provide accurate, representative information over the oceans on pressure, wind, sea-surface temperature, and rainfall frequency. Diurnal variations, except in pressure, are insignificant,

8

and long-term averaging of the readings of thousands of barometers has removed any bias in the pressure data.

Such assurance is unjustified over the continents, however, even with the most careful compilations (Braak, 1921–1929; Taylor, 1932; Biel, 1945; Ramanathan and Venkiteshwaran, 1948; Chu, 1962; Thompson, 1965; Lebedev and Sorochan, 1967) because of topographic and other distortions.

Reducing pressures observed over Tibet to mean sea level is meaningless, and over much of Africa is extremely difficult. The values computed by Weickmann (1963) for Africa are the best that are available but, even so, they must be treated with caution, especially over the Ethiopian highland. Moreover, diurnal bias resulting from averaging observations made at only a few hours each day is often difficult to detect.

Surface winds inland are susceptible to local distortions, while diurnal variation along the coast may completely mask a light mean wind, and, as Ananthakrisnan and Rao (1964) have shown for India, significant diurnal variations may occur hundreds of kilometers inland.

Over land, screen temperatures are liable to serious errors from wrong exposure and ventilation of thermometers, while the risk of rainfall measurements being unrepresentative is well known.

The reader, aware of caveats, can still obtain significant insight into monsoon regimes from studying these charts while remembering that the circulation patterns, as revealed by surface winds over the African plateau, may not conform too closely to the patterns of pressure reduced to mean sea level. Low-level monsoon flows originate where air temperature is lowest, in the centers of polar continental or subtropical anticyclones, and terminate in the centers of depressions which coincide with maximum air temperatures (heat lows) or maximum sea-surface temperatures.

2.2 Large-Scale Coastal Influences

The continental shore, coinciding with a zero-order discontinuity in heat capacity, is essential to the existence of monsoons. The coastline significantly influences the monsoons in two other respects: by being a material boundary to the ocean and by coinciding with a sharp first-order discontinuity in surface roughness.

2.2.1 SINKING AND UPWELLING

Ekman (1905) demonstrated that wind stress τ_0 transports water horizontally in a direction turned 90° anticyclonically from the direction of the stress. The rate of mass transport is given by τ_0/f (Reid, 1967). When the wind direction is parallel to a coast, mass continuity requires that the alongshore surface layers of the ocean must undergo compensating vertical motion. When the coast is on the anticyclonic side of the wind, sinking occurs; when the coast is on the cyclonic side of the wind, upwelling occurs. If the upwelling extends to the thermocline, it brings cold water to the surface.

2.2.2   STRESS-DIFFERENTIAL-INDUCED DIVERGENCE

Bryson and Kuhn (1961) noted that air flowing parallel to a coastline at the gradient level is subjected to more drag by the land than by the sea. As a consequence, surface winds will cross the isobars toward lower pressure at a greater angle over the land than over the sea. If the pressure inland is low, winds will diverge at the coast, but if the pressure inland is high, they will converge at the coast (Bergeron, 1949). Bryson and Kuhn, assuming no variation in drag parallel to the coast, expressed the horizontal divergence of the volume transport within the atmospheric friction layer as

$$[\Delta C_D/(f\,\Delta y)][(s\sin\alpha)^2 - (s\cos\alpha)^2] \tag{2.1}$$

where ΔC_D is the drag coefficient difference across the coast ($\approx 10^{-3}$), the x coordinate is the coast, and α is the angle between the surface wind and the coast.

What is the combined effect on monsoons of wind stress on the ocean and of the difference between over-land and over-ocean wind stress?

2.2.3   COASTAL INFLUENCES IN WINTER

Particularly over Southeast Asia, winds around a vigorous intensifying polar continental anticyclone blow at almost 90° across the isobars, which usually nearly parallel the coast. During late winter and early spring, however, continental anticyclones are often weak and then winds blow along the coast, transporting relatively warmer offshore water toward the coast. Coastal convergence resulting from the stress differential should lead to increased cloudiness and rainfall (Bergeron, 1949) reducing insolation at the land surface.* The inshore sea-surface temperature would thus be somewhat higher, the temperature gradient between sea and land somewhat larger, and the monsoon circulation somewhat stronger and more persistent than might otherwise obtain.

2.2.4   COASTAL INFLUENCES IN SUMMER

Winds around a continental heat low may blow almost parallel to a coastline, as off Somalia, for example. Cold water, being brought to the surface,

---

* Weather satellite pictures suggest the presence of this effect.

cools the adjacent air, producing a weak, local pressure ridge (Ramage, 1968a) and a large thermal gradient (Bunker, 1967). These two effects, by combining to reduce inflow to the heat low and to increase the winds, enhance upwelling. Thus feedback strengthens the monsoon. The stress differential leads to divergence which both enhances upwelling and causes lower-tropospheric subsidence, and which by clearing the skies, contributes in turn to insolation at the land surface. The combined effect is to increase the temperature gradient between land and sea and thereby to intensify the monsoon circulation.

Cold water may be upwelled or may be advected into the winter hemisphere where it may reduce the sea–land temperature gradient and, as subsequent paragraphs will show, may weaken or inhibit a winter monsoon.

2.3 Annual Variation of Surface Circulations

2.3.1 JANUARY MONSOONS*

2.3.1.1 *Southern Asia and the Northern Indian Ocean.* In winter, the low heat capacity of the land relative to the sea ensures that surface air over land is colder than that over the sea. A gradient of air pressure is therefore established, causing surface air to flow persistently from land to sea. Since the surface-pressure distribution changes little, a compensating flow must occur from sea to land in the higher troposphere. Over southern Asia, the Himalayas and the Iranian and Caucasus mountains block the southward movement of extremely cold air accumulated over central Asia. Thus the northeast monsoon is here a gentle phenomenon reflecting small temperature gradients between the land south of the mountains and the Indian Ocean. Over the central and eastern Indian Ocean the monsoon spirals into a weak low-pressure trough just north of the equator (not evident with 2-mb isobar spacing); over the western Indian Ocean the influence of Africa extends the monsoon across the equator.

2.3.1.2 *Eastern Africa and the Western Indian Ocean.* The African continent spans the equator and so in January radiational cooling results in high pressures over the Sahara and Arabia; radiational heating results in low pressure over the Kalahari Desert. The consequent north–south pressure gradient sets up a flow of air from north to south across the equator. The most intense heat lows overlie deserts and occupy the same latitude over the continents as do the subtropical anticyclones over the oceans. Even though *surface* flow

---

* See Figs. 2.1, 2.2, Plate I.

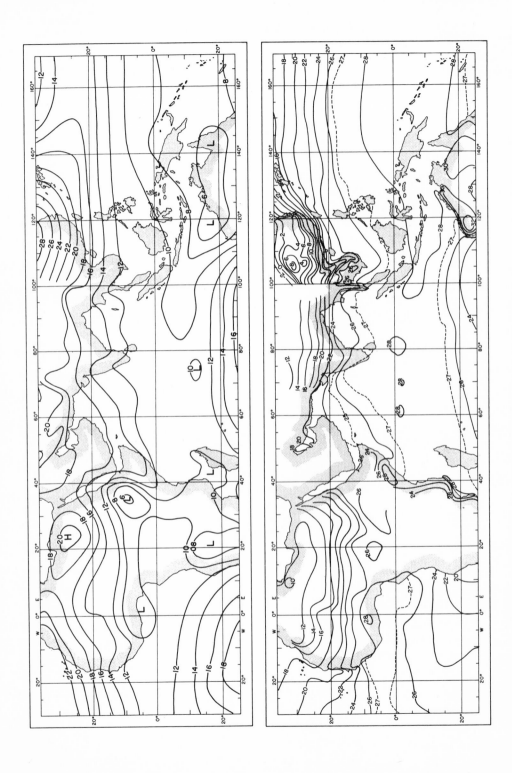

converges into the heat lows (Section 3.2.1), persistent subsidence through most of the troposphere precludes development of clouds.

The influence of Africa on the atmospheric circulation, persisting 800 km east of the continent and merging with the influence of Asia farther north, extends the northeast monsoon of the Arabian Sea far into the Southern Hemisphere.

2.3.1.3   *Southeast Asia and the China Seas.* Unobstructed by the Himalayas and channeled by mountain ranges east of Burma, the vigorous northeast monsoon of Southeast Asia blows out from the intense polar anticyclone centered near Lake Baikal, in a succession of surges often exceeding gale force. Cold water driven southward along the coast helps maintain temperature and pressure gradients between the continent and the China Seas that are twice as large as those between southern Asia and the Indian Ocean.

2.3.1.4   *Western Africa and the Eastern North Atlantic.* The anticyclone over the Sahara is the source of cold air streaming south but outbreaks of polar air from Europe are hindered by the relatively warm Mediterranean from intensifying the West African monsoon. The effect is not dissimilar to that produced farther east by the Himalayas, although, latitude for latitude, average temperatures are lower over Africa. In contrast to eastern Africa and the western Indian Ocean, the monsoon does not penetrate beyond the Guinea coast at the latitude of the warmest waters. Farther south, the South Atlantic anticyclone causes cold water to upwell in the Benguela Current (Hart and Currie, 1960) which then carries the cold water northward into the tropics. According to Böhnecke (1936), maximum upwelling occurs closest to the equator during the Southern Hemisphere summer. By cooling the surface-air layer, the current extends the anticyclone across the equator, and so prevents transequatorial flow from the Northern Hemisphere into the Kalahari heat low (Thompson, 1965). At latitude 15S, Atlantic sea-surface temperatures are almost 4C below those of the western Indian Ocean (see Serviço Meteorológico de Angola, 1955). Thus the Mediterranean Sea and the Benguela current combine to restrict the West African winter monsoon to 15° of latitude in the Northern Hemisphere while off eastern Africa the monsoon ranges over more than 40° of latitude.

2.3.1.5   *Indonesia.* The great mountainous islands of the "maritime continent" lying in a weak low-pressure trough in very warm seas are a

---

Fig. 2.1.   (*left*)   January: mean sea-level pressure (mb) (in tens and units).

Fig. 2.2.   (*right*)   January: mean surface temperature (°C) in the screen (over land) and of the sea surface (over the ocean). No attempt has been made to extend the analysis over high ground.

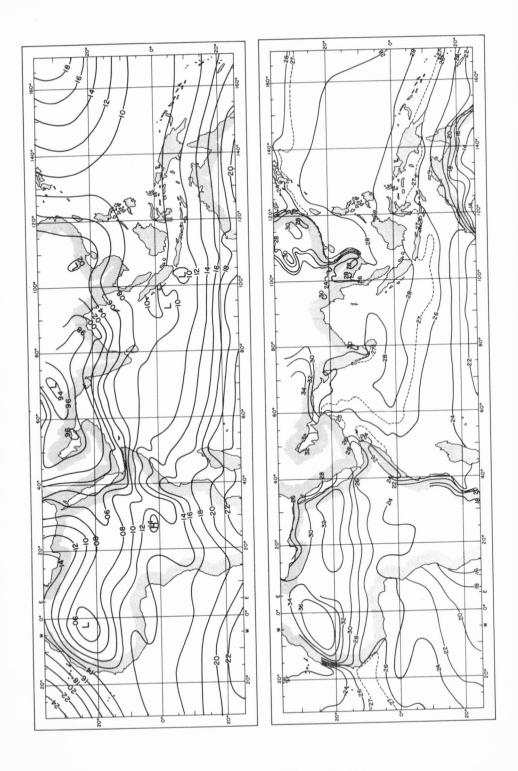

convergent focus for air streaming south in the vigorous Southeast Asian winter monsoon.

2.3.1.6 *Australia*. Cyclonic circulation about a desert heat low over northern Australia comprises a limited summer monsoon which extends its influence eastward as far as the Solomon Islands, and is strongest in the west where land–sea temperature gradients are large.

In the mean a continuous pressure gradient extends from southeastern Asia to northern Australia. However, it is nearly flat over Indonesia, and so, compared to the western Indian Ocean, relatively little air moves across the equator between the northern winter and the southern summer monsoons (Section 7.2.1).

Fresh, steady southerly winds blow parallel to the coast of western Australia, but no upwelling has been detected (Wooster and Reid, 1963). Wyrtki (1969) suggested that upwelling does in fact occur but that the upwelled water is replaced by *warmer* subsurface water drifting southward around Northwest Cape. This water in turn upwells but its effect is only to *raise* slightly the sea-surface temperature along the coast. The upwelling warm water should therefore weaken the Australian summer monsoon by reducing the land–sea temperature gradient.

2.3.2 JULY MONSOONS*

2.3.2.1 *Southern Asia and the Northern Indian Ocean*. As in winter, the Himalayas and the Iranian and Caucasus mountains seal off regions to the south from invasions of cold air. Thus over Arabia, southern Iran, West Pakistan, and northern India an intense heat low dominates the circulation and fresh or strong southwest winds prevail to the south. Whereas cold up-welled water inhibits development of a summer monsoon off southwestern Africa, off the coasts of Somalia and Arabia (Swallow, 1965; Stommel and Wooster, 1965) it intensifies the southwest monsoon, by accentuating the summertime reversal of the land–sea temperature gradient.

2.3.2.2 *Eastern Africa and the Western Indian Ocean*. In contrast to January, radiational cooling results in high pressure over the Kalahari Desert, and

---

* See Figs. 2.3, 2.4, and Plate III.

Fig. 2.3. (*left*) July: mean sea-level pressure (mb) (in tens and units).

Fig. 2.4. (*right*) July: mean surface temperature (°C) in the screen (over land) and of the sea surface (over the ocean). No attempt has been made to extend the analysis over high ground.

radiational heating, in low pressure over the Sahara. The south–north pressure gradient sets up a southerly flow across the equator, eventually merging with the southeast trades over the southern Indian Ocean and with the southwest monsoon north of the equator. The upwelling effect, mentioned above contributes to the southerly monsoon being stronger than the northerly monsoon of January.

2.3.2.3 *Southeast Asia and the China Seas.* Cool invasions from the north (Chu, 1962) and tropical cyclones from the east ensure that a vigorous heat low cannot develop east of the Himalayas. Thus, although air over land is warmer than air over sea in July, the pressure gradient is small and the southwest monsoon is much weaker than over the northern Indian Ocean.

2.3.2.4 *Western Africa and the Eastern North Atlantic.* A rather large pressure gradient between the Saharan heat low and the South Atlantic anticyclone results in vigorous southerly flow across the equator. However, because of the year-round persistence of cold water south of the equator, and because the land between 10 and 20S remains warmer than the sea and the seat of a heat low (Weickmann, 1963; Flohn and Strüning, 1969), only that portion of the flow north of 5N can be classed as monsoonal.

2.3.2.5 *Indonesia and Australia.* In July Indonesia occupies the same position relative to a mean polar high over Australia as it occupies relative to the Siberian high in January. In contrast to the persistent Siberian winter high, high pressure over Australia is but the climatological average of eastward-migrating anticyclones which may temporarily slow and intensify over the relatively cold continental interior (Russell, 1893). Some transfer takes place across the equator into the Southeast Asian heat low.

2.3.3 MONSOONLESS SOUTH AMERICA

From this discussion it is possible to see why South America is not monsoonal. The continent narrows away from the equator both to north and south, thus restricting the areas in which polar highs or heat lows might form. In addition, in the Northern Hemisphere the continent does not extend into the latitudes of general subsidence characterized by the subtropical highs over the oceans and favoring the heat lows of northern Africa and southern Asia. Finally, except for brief interruptions, persistent upwelling along the west coast, induced by the southeast trades and *reaching a maximum in winter* (Wooster and Reid, 1963), maintains sea-surface temperatures that are below overland surface-air temperatures throughout the year.

2.4   Heat Balance at the Ocean Surface during the Monsoons

Wind stress on the ocean surface induces currents which can lead to pro-found effects near the shore (Section 2.2). Over the ocean as a whole, inter-action at the air–water interface importantly affects both gaseous and liquid realms. Discussion in this section is confined to large-scale heat exchange associated with the monsoons. January and July mean patterns are described and compared (Fig. 2.5). In Chapter 5, I relate annual variations at typical ocean locations to the march of the seasons (Fig. 5.7).

2.4.1   DETERMINATION OF HEAT EXCHANGE

Many investigators have tried to derive empirical equations relating the elements of standard marine meteorological observations to the heat budget at the ocean's surface. All the equations suffer from a lack of accurate direct measurements of the budget components, in particular during periods of strong winds and high seas. The equations used here are based chiefly on the work of Budyko (1956) and Roden (1959). Although the patterns of heat exchange derived from the equations are probably realistic, absolute values and magnitudes of gradients qualify only as approximations (Garstang, 1965).

In the heat-balance equation (in langleys per day):

$$Q_{st} = Q_s - Q_r - Q_b - Q_e - Q_h \tag{2.2}$$

where*

$$Q_s = Q_0(1 - \bar{N}k) \tag{2.3}$$

$$Q_r = (Q_s \times k') \tag{2.4}$$

$$Q_b = S\sigma T_s^{4}(0.39 - 0.05e_a^{-2})(1 - k''\bar{N}^2) + 4S\sigma T_s^{3}(T_s - T_a) \tag{2.5}$$

$$Q_e = 5.17(e_s - e_a)V, \qquad \text{where } e_s > e_a \tag{2.6a}$$

$$Q_e = 7.7 \times 10^{-5}(444 - 0.56T_s)(0.98e_s - e_a)V, \qquad \text{where } e_s < e_a \tag{2.6b}$$

$$Q_h = Q_e(0.66)(T_s - T_a)/(e_s - e_a) \tag{2.7}$$

$Q_s - Q_r - Q_b = Q_{eff}$, the net or effective radiation at the sea surface, which, if positive, heats the surface and, if negative, cools the surface.

---

* As is true in many empirical formulations, the equations are not dimensionally balanced.

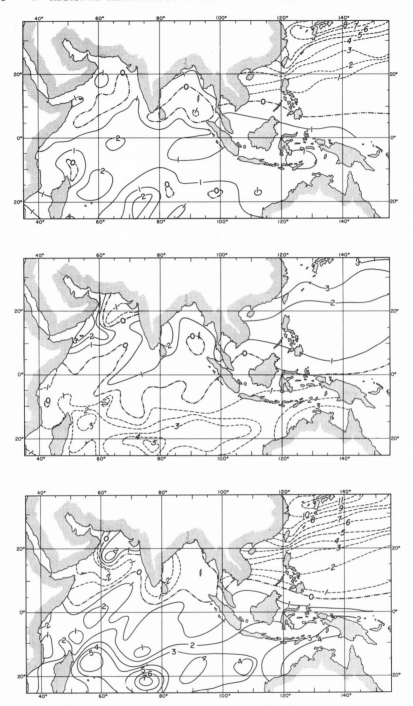

Fig. 2.5. Net heat balance at the ocean surface in hectolangleys day $^{-1}$ (Ramage, 1969b; Wyrtki, 1957, 1966). (top) January; (middle) July; (bottom) January minus July.

The amount of solar radiation (Q_s) reaching the sea surface depends on the season and cloudiness, whereas the amount of heat abstracted from the sea by the air is directly related to the wind speed and the air–sea temperature difference, in the case of sensible heat (Q_h), and to the wind speed and the vertical gradient of vapor pressure, in the case of latent heat (Q_e). In the tropics Q_{eff} and Q_e are of comparable magnitude and oppositely directed, and are an order of magnitude larger than Q_h.

2.4.2   HEAT BALANCE IN JANUARY*

In the Northern Hemisphere, solar radiation is least in January. However, the sea gains about one hectolangley per day over the northern parts of the Arabian Sea and Bay of Bengal where clear skies scarcely interrupt the sun's radiation and light winds cause little evaporative cooling. Farther south, more cloud and fresher winds counteract greater solar radiation incident at the top of the atmosphere to produce a slight net cooling at the sea surface.

Over the China Seas a combination of clouds, reducing insolation, and strong cold dry winds, enhancing evaporative and sensible cooling, rapidly removes heat from the sea (chiefly in latent form). The largest gradients of net heat balance coincide with the northwestern boundary of the warm Kuroshio Current. The heat, eventually released in heavy rain over Indonesia, helps drive the vigorous East Asian Hadley cell (Section 5.8.2).

In the equatorial zone and off southeastern Africa and northern Australia, the ocean gains 1–2 hectolangleys per day. To the east, considerable cloudiness tends to reduce the effect of only slight evaporative cooling from light, moist winds. To the west, fresh winds combine with partly cloudy skies to produce about the same net effect.

2.4.3   HEAT BALANCE IN JULY†

As in January, patterns of net heat balance reflect considerable differences between the South Asian and the East Asian monsoon regimes.

The surface of the western Arabian Sea gains most heat in the region of upwelling. The few clouds scarcely interrupt solar radiation; the sea gains both sensible and condensation heat from the air, for not only is $T_s < T_a$ but $e_s < e_a$ (Fig. 5.41). Thus the monsoon, by bringing cold water upwelling to the surface, ensures that the water column is heated. Downwind beyond the upwelling zone, strong winds, previously cooled by the upwelled water, cause evaporative cooling intense enough to overcome midsummer insolation; off northeastern Arabia net heat balance gradients exceed one hectolangley

---

* See Fig. 2.5, top.
† See Fig. 2.5, middle.

$100 \, \mathrm{km}^{-1} \, \mathrm{day}^{-1}$. Patterns over the Bay of Bengal weakly resemble those over the Arabian Sea.

Although cloudiness reduces insolation, the surface of the China Seas gains heat, because light moist winds result in only slight evaporation. Off southeastern Africa and northern Australia evaporative cooling, through the agency of fresh southeasterlies, overcomes insolation.

2.4.4 DIFFERENCES BETWEEN SUMMER AND WINTER*

Off eastern Asia and northern Australia the ocean surface loses heat in winter and gains heat in summer. The much greater difference over the China Seas than to the south stems from the much stronger East Asian winter monsoon.

Summer and winter net heat balances differ little over the eastern parts of the Arabian Sea and Bay of Bengal, while over the northeastern Arabian Sea the surface is heated more in winter than in summer. In summer, as compared to winter, stronger winds enhance evaporative cooling and greater cloudiness reduces insolation. The similar though less marked differences off southeastern Africa can be similarly explained.

In the monsoon area annual variation of the net heat balance east of 100E differs sharply from the annual variation west of 100E, reflecting major discontinuities in the annual variations of all meteorological elements. The cause already suggested (Section 2.3) and further discussed in Section 5.1, is orographic—the mountain ranges of southern Asia and the Tibetan Plateau.

The sea surface in equatorial regions gains more heat in January than in July. Over the Indian Ocean it is brought about by the northward extension of fresh winds during the Southern Hemisphere summer. No ready explanation for the difference east of 100E comes to mind.

2.5 Rainfall Distribution

Plates I and III reveal no simple relationship between monsoon rainfall and surface monsoon flow. Since much of the remainder of this book will be concerned with this problem, only some general comments need be made now.

Anticyclones and ridges are dry; the heat lows are dry, and even over the oceans the main low pressure troughs are relatively dry.

In the monsoon area most rain falls in summer or in autumn, except for a near-equatorial band possessing a double maximum (Fig. 2.6). The summer

---

* See Fig. 2.5, bottom.

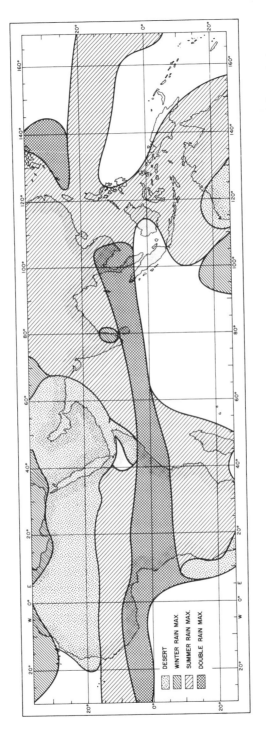

Fig. 2.6. Rainfall distribution through the year. Periods during which more than 75% of the annual rain falls are areally delineated. Areas with less than 250 mm yr^{-1} are classed as deserts; areas accumulating 75% of the annual rainfall in over 7 months are considered to have no seasonal maximum. Ship observations of rainfall frequency are tied in to the land station data. Small areas with sharp orographic distortions have not been identified.

maximum is associated with heat-low intensification, although the heat lows themselves are dry (Section 3.2). The double maximum is probably associated with temporary intensification of near-equatorial troughs in the transition seasons (Fig. 5.18).

Despite the fact that moving surface disturbances are rare in the monsoon region and that surface winds exhibit notable steadiness during summer and winter, rainfall departs as much from normal as in other regions (Fig. 2.7).

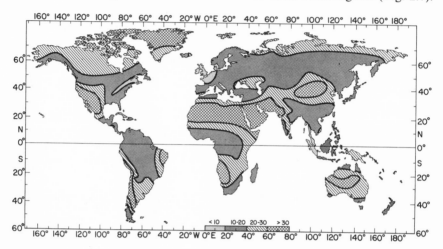

Fig. 2.7. Average departure from normal of annual precipitation, according to Biel (1929), depicted as a percentage of the average precipitation.

The rather peculiar causes of this variability are treated in subsequent chapters.

2.5.1   PERIOD AND SCALE RANGES OF MONSOON RAINFALL VARIATIONS

The familiar annual period acts within a narrow compass over western Africa as compared with southern Asia and the Indian Ocean.

Interannual variability is high (Fig. 2.7). Thus far no one has detected a predominant period. Some examples are considered in more detail in Chapter 7.

Within a single monsoon season, at least five types of variation, two of them climatological, interact. Over Southeast Asia, the character of the winter monsoon changes during January (Section 5.8.2.6), while a dry spell in July (Fig. 5.38) separates the early and late parts of the summer monsoon (Chapter 5).

Diurnal variations (Section 4.1.3) range in scale from local mountain-valley circulations through sea breezes to subcontinental systems (Ananthakrishnan and Rao, 1964). For example, along the coast, the latter two may be out of phase, interacting to produce a complex local diurnal variation (Nicholson, 1970).

Synoptic changes are effected by moving disturbances (Sections 3.3, 3.4, and 3.11) and by intensification and decay of quasi-stationary bad-weather

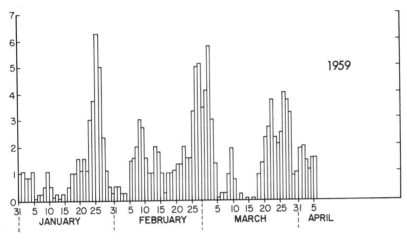

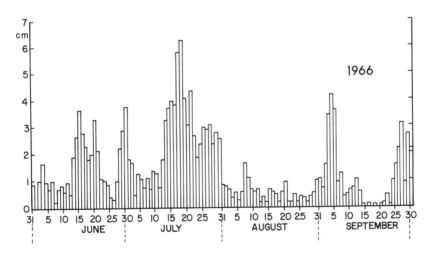

Fig. 2.8. (Top) daily rainfall "indexes" for Kenya (Johnson, 1962); abscissa, tens of percent of stations reporting rain.   (Bottom) average daily rainfall along the west coast of peninsular India (Ananthakrishnan, 1966).

systems (Sections 3.5, 3.8, 3.9, and 3.10). Resultant weather is locally modified by the diurnal cycle and by mesoscale distortions (Rainbird, 1968) (Chapter 4) and is also constrained within the envelopes of important more extensive fluctuations with periods similar to those of the middle-latitude Grosswetter-lagen (Baur, 1951). Figure 2.8 clearly shows existence of synoptic and Gross-wetterlagen periods over India (Ananthakrishnan and Keshavamurthy, 1970) and eastern Africa. Their interaction is a recurrent theme.

2.5.2   TYPES OF MONSOON PRECIPITATION

In the absence of orographic shadowing, most rain falls in one of these two circumstances: either from

(1) deep nimbostratus with embedded cumulonimbus when vertical wind shear and lower tropospheric convergence both are large (although rain intensity may fluctuate considerably, skies remain predominantly overcast), or

(2) scattered towering cumulus or cumulonimbus, when vertical wind shear and lower tropospheric convergence both are small (see Braak, 1921–1929).

I shall assign the term *rains* to the former and *showers* to the latter, realizing of course that sharp demarcation is impossible.

When these westward Winds are thus settled, the Sky is all in mourning, being covered with black Clouds, pouring down excessive Rains sometimes mixt with Thunder and Lightning, that nothing can be more dismal.

William Dampier, *A New Voyage round the World*, 1697.

3. Synoptic Components of Monsoons

By definition (Chapter 1), alternations of surface cyclones and anticyclones are rare in the monsoon area. Nevertheless, variability of annual rainfall is as high there as it is in areas where such alternations are frequent, and where, presumably, annual rainfall variability might also be high (Fig. 2.7).

It is the purpose of this chapter to describe the most common components of the monsoons and hence to open the way for detailed discussion of how they contribute to the surprisingly high rainfall variability.

The components are conveniently subsumed within three genera:

(a)   Circulations in which downward motion and fine weather predominate, i.e., polar and subtropical anticyclones, and heat lows.

(b)   Circulations (generally summertime) in which upward motion and wet weather predominate, i.e., tropical cyclones, monsoon depressions, and subtropical cyclones.

(c)   Other systems with marked weather gradients, i.e., troughs in the upper tropospheric westerlies, near-equatorial troughs, quasi-stationary non-circulating disturbances, surface transequatorial flow, and squall lines.

Within the first two genera, species overlap, and cannot be sharply demarcated. Figure 3.1 arranges the species according to divergence and vertical motion.

The relationship of divergence and vertical motion is discussed in Section 6.1.5. Errors in measurement and lack of representativeness preclude accurate computation of *synoptic* divergence, and even of the sign of the accompanying vertical motion. Averaging, however, usually reduces random errors sufficiently to enable mean vertical motions to be estimated, and it is on these means that Fig. 3.1 is based.

25

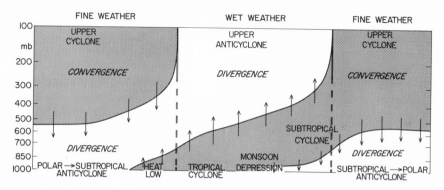

Fig. 3.1. Circulation components of the monsoons arranged schematically according to weather, divergence, and vertical motion. Levels of nondivergence are denoted by heavy lines.

Averaging may be performed in more than one way to derive models of monsoon components. In the case of tropical cyclones (Section 3.3.2), observations made in great numbers of storms are averaged on a polar coordinate system in which the origin coincides with the storm center and the angular coordinate is measured clockwise from the direction of storm movement (Fig. 3.7).

A subtropical cyclone model (Section 3.5.3) was described by compositing the meteorological fields over several days around a circulation which neither moved nor changed intensity much during that period.

The major monsoon circulations are often strong and persistent. Thus divergence computed from monthly mean resultant winds can be used to determine whether the sense of the predominating mean vertical motion is up or down. Regions of mean rising motion are also regions of relatively greater mean cloudiness and greater mean rainfall; conversely mean sinking motion occurs in regions where weather is relatively good.

In Fig. 3.1, upper-tropospheric divergence is associated with wet weather and upper-tropospheric convergence with fine weather whereas the converse is not necessarily true *for the lowest layers*. Within each genus, species shade into one another. A rare subtropical cyclone or monsoon depression may develop into a tropical cyclone. West African depressions which are best developed at 850 mb (Gilchrist, 1960) and weak disturbances over northern Australia (Southern, 1969), resemble both subtropical cyclones and monsoon depressions. Over the desert, a heat low replaces an anticyclone as spring gives way to summer. In the western Pacific an upper tropospheric depression sometimes extends downward (Palmer, 1951) and, changing from genus 1 to 2, passes through the subtropical cyclone stage to typhoon (Ramage, 1959).

Here the attentive reader has probably already remarked on the absence of the two well-known terms—Intertropical Front and Intertropical Convergence Zone. Northern Hemisphere and Southern Hemisphere air masses often meet along the axis of the heat trough. Over the continents, the air mass originating in the summer hemisphere is tropical continental and the air mass originating in the winter hemisphere is usually tropical maritime (see Plates I and III). Thus there is some justification in referring to the meeting zone as the Intertropical Front or ITF. Unfortunately, as is discussed in Sections 3.2, 3.8, and 5.3, the ITF and the worst weather do not coincide (Thomspon, 1965; Genève, 1966; Garnier, 1967).

Tropical meteorologists who were confused by this anomaly compounded their confusion by applying a new term, Intertropical Convergence Zone (ITCZ), to the bad weather—"convergence zone," yes, "intertropical," no. Confusion became chaos when the existence was confirmed of troughs and convergence zones near the equator in the *winter* hemisphere. Valiant traditionalists persisted in applying the adjective "intertropical" to simultaneously existing troughs or convergence zones on both sides of the equator while others used ITF and ITCZ indiscriminately and interchangeably. This book has no place for such jargon aphasia; I trust that heat low, heat trough, near-equatorial trough, or near-equatorial convergence will prove to be orderly and intelligible alternatives.

Tropical cyclones, anticyclones, and troughs in the upper tropospheric westerlies are briefly mentioned since they have been extensively described in other texts; heat lows and quasi-stationary noncirculating disturbances receive similar treatment but for a different reason—scantiness of data.

3.1 Polar Anticyclones

Polar anticyclones have not attracted much attention recently. As Wexler (1951), succinctly summarized:

"The polar anticyclone is created by cooling of the surface layer of air, which loses its heat to an underlying cold surface by radiation or by eddy conductivity. The lower temperature of the surface itself is caused by the nocturnal radiational cooling of snow and ice fields, such as those in polar regions, or by bodies of larger thermal inertia (heat capacity) such as oceans and large lakes. The cooling and vertical shrinking of the surface layer of air depresses the isobaric surfaces aloft, creating an intensifying 'polar anticyclone' which causes an inflow of air across the isobars, thus creating a larger barometric pressure at the surface."

During winter, the Siberian polar anticyclone, though centered far to the north, dominates Southeast Asia. Much weaker winter polar anticyclones persist over Libya and Egypt while eastward-moving anticyclones temporarily slow and intensify over Australia.

Two influences, one in the upper troposphere, the other in the surface layers, almost always combine to favor development of a vigorous polar continental anticyclone.

3.1.1   UPPER TROPOSPHERE

In the vorticity equation,

$$\nabla \cdot \mathbf{V} = -(\zeta + f)^{-1}(d\zeta/dt + \beta v) \tag{3.1}$$

$$d\zeta/dt = \mathbf{V} \cdot \nabla\zeta + w\, \partial\zeta/\partial z + \partial\zeta/\partial t \tag{3.2}$$

In the high troposphere, vorticity is usually a maximum. Therefore $\partial\zeta/\partial z$ is small and since w is at least two orders of magnitude smaller than V, $w\, \partial\zeta/\partial z$ can be neglected. Since air moves rapidly through an upper-tropospheric westerly wave system, $\partial\zeta/\partial t$ is also small. Thus the sign and magnitude of $d\zeta/dt$ is determined by $\mathbf{V} \cdot \nabla\zeta$, the horizontal advection of vorticity.

Vigorous polar anticyclones generally develop beneath and east of upper-tropospheric ridges, where convergence is found. This convergence results from positive values of $\mathbf{V} \cdot \nabla\zeta$ (advection of negative vorticity) exceeding negative values of βv, while since f almost always exceeds ζ (which is generally negative), $f + \zeta > 0$.

3.1.2   LOWER TROPOSPHERE

Petterssen (1956) enlarging on Sutcliffe's (1947) work developed an equation*

$\partial\zeta_0/\partial t$

$$= A_\zeta - (R/f)\nabla^2\{\log(p_0/p)[c_p^{-1}(d\overline{H}/dt) + \langle\omega(\Gamma_a - \Gamma)\rangle_{av}] + (g/R)A_T\} \tag{3.3}$$

The contribution of the first term on the right-hand side has been already discussed in the preceding section. Accurate determination of A_ζ demands positioning of the level of nondivergence, which may vary considerably in space and time. Hence in the preceding section I chose to use the vorticity

---

* Here, $p_0 = 1000$ mb and p is taken to be the pressure at the level of nondivergence; the bar and "av" signify mean values in the layer between p and p_0.

equation to determine divergence in the upper troposphere and the consequent effect on the mass of the tropospheric column. A_ζ is approximately
proportional to upper tropospheric values of $-\mathbf{V} \cdot \nabla \zeta$ and $\nabla \cdot \mathbf{V}$.

The second term,

$$-(R/f)\nabla^2[\log(p_0/p)c_p^{-1}(\partial \bar{H}/dt)]$$

represents diabatic influences. Over a cold source, such as Siberia, since
$\partial \bar{H}/dt < 0$ and increases outward, $\nabla^2 \partial \bar{H}/dt > 0$.   Thus the cooling effect
mentioned by Wexler favors anticyclogenesis.

The third term,

$$-(R/f)\nabla^2[\log(p_0/p)\langle \omega(\Gamma_a - \Gamma)\rangle_{av}]$$

represents the effect of adiabatic changes. In an anticyclone, subsidence
prevails. Consequently, $\omega > 0$, being downward, $(\Gamma_a - \Gamma) > 0$, and both
probably decrease outward from the anticyclone center. Therefore, ∇^2
$\langle \omega(\Gamma_a - \Gamma)\rangle_{av} < 0$. Since subsidence in the lowest layers *results from* anticyclogenesis, this term acts as an effective inhibitor only after the pressure
gradient has increased sufficiently for marked outflow to develop. It then acts
against the other terms to limit the intensity of the anticyclone.

The fourth term,

$$-(g/f)\nabla^2 A_T$$

represents thickness advection. Near the center of a polar anticyclone, where
temperatures are lowest, this term opposes the second term since $A_T > 0$ and
$\nabla^2 A_T < 0$. According to Petterssen, the term contributes to increasing anticyclonic vorticity west of a developed cyclone and thus will usually act in the
same sense as the advection of upper-tropospheric vorticity.

The basin of Lake Baikal in Siberia is the seat of the most intense polar
high. Mountain barriers immediately to south and east, by retarding outflow,
reduce the inhibiting effect of subsidence (third term).

In his numerical modeling of the global atmospheric circulation, Mintz
(1968) found presence of the Himalayan–Tibetan massif to be essential if the
model were to reproduce the winter Siberian anticyclone. Without the mountains, convectively heated air from the Indian Ocean could move northward
and counteract the effect of radiative cooling (second term). Thus the mountains significantly diminish the role of thickness advection (fourth term) in
anticyclolysis.

Upper-tropospheric convergence with surface cooling combine to raise
surface pressure. The pressure gradient may increase abruptly and outflow

from the high may accelerate. Almost always when upper-tropospheric waves are intense and of large amplitude, surface cyclogenesis occurs east of the upper trough, particularly off eastern Asia where surface temperature and moisture fields favor frontogenesis. Cyclogenesis, through thermal advection, further intensifies the polar high to the west. Consequently, surface cyclogenesis usually precedes or accompanies intensification of the polar high and the leading edge of the equatorward surge of cold air is part of the cold front extending equatorward and westward from the cyclone. However, now and then over eastern Asia upper convergence and surface cooling may so intensify the polar high as to initiate a surge without surface cyclogenesis farther east (Bell, 1968).

3.1.3   EXAMPLES OF POLAR ANTICYCLONES

3.1.3.1   *Intense Anticyclogenesis over Eastern Asia.* Occasionally super-anticyclones develop over Southeast Asia, and perhaps over northern Africa (Section 5.3.2.1). Danielsen and Ho (1969) used isentropic trajectories to analyze intensification of a Siberian high to 1071 mb (see also Section 5.8.2.3).

In January 1967 the winter monsoon was particularly vigorous. Three surges crossed Hong Kong, where average pressure for the month was 2.3 mb above normal (Fig. 5.20).

On the 11th, during a minor lull, a ridge covered South China with pressures ranging from 1025 to 1035 mb. In the north a short wave trough in the polar westerlies, with an accompanying surface low, was beginning to move southeastward to the northeast of Lake Baikal. Farther west near 70E, an intensifying upper tropospheric ridge extended to the Arctic Ocean. Cold air was advected southward from the ridge into the trough by supergeostrophic winds and the trough intensified through energy dispersion (Yeh, 1949).

Although the air was descending (Fig. 3.2), cold advection was occurring on *isentropic* surfaces, i.e.,

$$-\nabla^2 \{\log(p_0/p)\langle \omega(\Gamma_a - \Gamma)\rangle_{av} + (g/R)A_T\} < 0.$$

In fact, cold advection was greatest *ahead* of the trough, where the flow pattern was strongly difluent. The subsidence and cold advection propagated anticyclogenesis downward, and, because of anticyclonic turning in the difluent field, southward as well. West of the trough, all the terms in Eq. (3.3), except the overpowered adiabatic term, contributed to anticyclogenesis.

Central pressure in the high rose to 1071 mb by the 13th and persisted at that level until the 15th, despite the fact that a massive surge had already developed across the South China Sea (Section 5.8.2.3). The trough in the

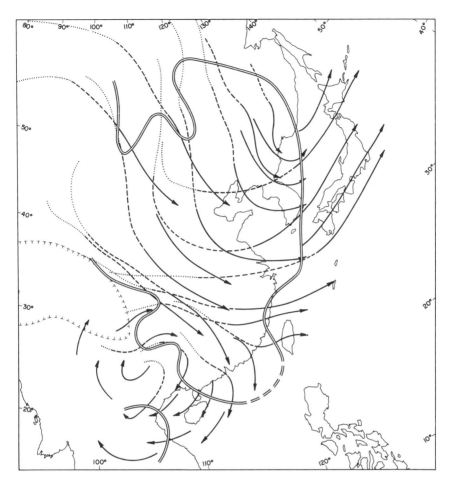

Fig. 3.2. 36-hr trajectories on the 300K isentropic surface, 14–15 January 1967: 0000 to 1200 GMT 14th, dotted lines; 1200 GMT 14th to 0000 GMT 15th, dashed lines; 0000 to 1200 GMT 15th, full lines. The zero vertical-motion isopleth (double line) encloses the region of subsidence.

The isentropic surface intersects the 500-mb pressure surface along about 30N and the 850-mb pressure surface near the coast of South China (Danielsen and Ho, 1969).

polar westerlies moved eastward. By the 16th it had reached the normal long-wave trough longitude near 125–130E (Fig. 5.2) and the monsoon began to weaken (Fig. 5.20).

The depth and extent of subsidence in this outbreak (Fig. 3.2) overpowered the normally persistent jet-stream divergence over central China. On the 16th (Fig. 5.21) China was clear of clouds except for the Red Basin of Szechwan

(Section 5.8.2.1). The 300K isentropic surface intersected the 700-mb pressure surface over southern China. The circulation there closely resembled those Thompson (1951) described as accompanying fine-weather monsoon surges.

According to Chu (1962), this situation typifies the most intense Southeast Asian cold surges.

3.1.3.2 *Features of an Australian Winter Anticyclone* are shown in a NNW–SSE cross section across the continent from Darwin (12°26′S, 130°52′E) to Mt. Gambier (36°45′S, 140°47′E) for 0000 GMT 2 August 1964 (Fig. 3.3). The cross section almost coincides with the major axis of the high. Two centers of 1030 mb were located at about 25 and 32S.

At 200 mb moderate convergence over the high center gave way to divergence farther north (Table 3.1). Surface temperatures were lowest near Alice

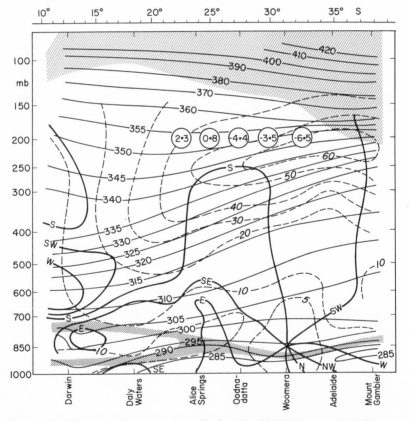

Fig. 3.3. Meridional section across Australia for 0000 GMT, 2 August 1964. Divergence in units of 10^{-5} sec^{-1} is entered in circles at 200 mb. Isogons are shown thick and full. The stratosphere is hatched and subsidence inversion layers are stippled.

Table 3.1

Divergence at 200 mb along the axis of an Australian polar anticyclone 0000 GMT 2 August 1964[a]

Lat. (°S)	V (deg)	V (m sec⁻¹)	ζ ($\times 10^{-4}$ sec⁻¹)	f ($\times 10^{-4}$ sec⁻¹)	β ($\times 10^{-10}$ m⁻¹ sec⁻¹)	$\mathbf{V}\cdot\nabla\zeta$, ($\times 10^{-8}$ sec⁻²)	βv, ($\times 10^{-8}$ sec⁻²)	$\mathbf{V}\cdot\mathbf{V}$, ($\times 10^{-5}$ sec⁻¹)
32.5	197	66	−0.27	0.76	−0.19	0.44	−0.12	−6.5
30	195	57.5	−0.28	0.73	−0.20	0.27	−0.11	−3.5
27.5	189	49.5	−0.25	0.68	−0.20	0.29	−0.10	−4.4
25	188	45	−0.11	0.62	−0.21	0.05	−0.09	+0.8
22.5	188	42	−0.11	0.55	−0.21	−0.01	−0.09	+2.3

[a] See Fig. 3.3.

Springs (23°48′S, 133°53′E) so that the cooling effect also favored anticyclogenesis. However, widespread subsidence prevailed and although the subsiding air did not appear to reach the surface south of 15S, nevertheless the adiabatic term must have opposed anticyclogenesis. Between the surface and 500 mb, weak cold advection over the high contributed to intensification (chart not shown). At the time we are considering here the high had reached its greatest intensity; within 24 hr central pressure fell 2 mb.

3.2 Heat Lows

3.2.1 DESCRIPTION

Well-developed heat lows are found only where the sun's summer rays reach the ground day after day through nearly cloudless skies. Absorption by air of reradiation from the ground reduces the density of the lower troposphere to create the heat low. Although the low weakens at night from radiational cooling (Fig. 3.4), it is a persistent, though shallow, summer feature.

Heat lows are predominantly desert phenomena, and for that reason have been poorly observed. The weather stays clear, and hostile. Over eastern Sudan and the Great Indian Desert, meager networks of aerological stations have been established, but only on the edges of the Saharan and West Pakistanian heat lows. Nevertheless, the lows are so persistent and vigorous that acceptable deductions can be derived on their characteristics, even from scanty data.

In a heat low, surface air converges toward the pressure minimum (Fig. 3.23) and rises. Only two inhibitors can then prevent significant cloud formation:

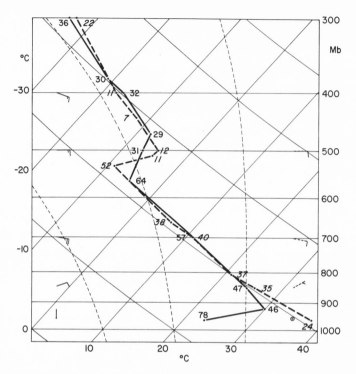

Fig. 3.4. Aerological soundings with relative humidity and winds (one full barb denotes 5 m sec^{-1}) for Khartoum (15°36′N, 32°33′E) at 0500 GMT (full line) and at 1100 GMT (dashed line), 23 September 1957.

extremely low humidity, a well-marked inversion, or a combination of both. At least one must be operating; otherwise cloud and rain, by interrupting the solar rays, would lead to a significant weakening of the heat low, a fact W. L. Dallas emphasized as long ago as 1887. It is no accident that heat lows are found close to the latitudes occupied by the oceanic subtropical highs (Figs. 2.1 and 2.3) and are affected by the same gentle mid tropospheric subsidence that keeps weather fine in the highs. The weather remains fine throughout the year under the influence of this subsidence; whether the surface circulation is a polar high (as in winter), or a summer heat low, seems to make little difference above 800 mb (Fig. 5.4).

Lower-tropospheric air generally converges into heat lows from continent and ocean (Plates I–IV) and thus along the axis of the heat trough (in which the heat lows lie) a considerable discontinuity may exist between dry desert air and relatively moist, originally maritime air. Over northern Africa the discontinuity is termed the "intertropical front." It is said to have a slope, with relatively cool moist "monsoon" air underlying relatively hot dry "harmattan" air. The frontal concept, however, is hard to sustain, for it does not coincide with a weather discontinuity (Bhalotra, 1963; Johnson, 1964d; Thompson, 1965). Nevière (1959) tried to show that this moisture discontinuity shifts diurnally across stations located nearby, moving to the north at night and to the south during the day. He admitted that the surface winds do not always shift appropriately to southerly directions at night and northerly directions during the day but pointed out that surface relative humidity satisfied his criterion for an air mass change ($<25\%$ for harmattan; $>50\%$ for monsoon).

Figure 3.4 shows the Khartoum soundings for 0500 and 1100 GMT, 23 September 1957, a day on which the heat trough lay close to the station and weather remained fine. By Nevière's criterion Khartoum lay in monsoon air at 0500 hr and in harmattan air at 1100 hr. A much simpler explanation is merely that throughout the period the station was in monsoon air, topped by a subsidence inversion at around 550 mb. All that happened was that rapid morning heating destroyed a surface radiation inversion which had developed during the night. Upper winds remained essentially unchanged.

3.2.2   EFFECT OF HIMALAYA–TIBET

Southern and Southeast Asia east of 70E are wet in summer and, except for brief periods,* free from heat lows. The pattern of upper-tropospheric mean winds is consistent—convergence above the heat lows (Plates I–IV), divergence over the rainy regions—but what creates the pattern?

Flohn (1960a) hypothesized that early summer warming of the Tibetan Plateau, by developing and increasing a north–south temperature gradient over southern Asia, causes upper-tropospheric easterlies to strengthen (Plate III), and encourages onset of summer monsoon rains. This is discussed more fully in Section 5.1, but how might the heat lows be affected? Figure 5.3 shows that at Lhasa in southern Tibet, mean monthly 300-mb pressure heights during summer are greater than at stations to east and west that are not on the plateau.

The consequently stronger easterlies south of Tibet would account for upper tropospheric convergence downstream above the heat lows and deserts, and divergence upstream in the region of heavy summer rains.

3.2.3   DYNAMIC AND THERMODYNAMIC FEATURES

The heat low is a persistent circulation which probably exports cyclonic vorticity in the middle and upper tropospheres (Section 3.5.4.1). Therefore,

---

* During spring a heat low occasionally develops over the Red Basin of Szechwan and usually persists for 3 or 4 days (Li, 1965) (Section 5.8.3.2).

the right-hand side of Petterssen's development equation (3.3) should be positive:

(*i*) *Advection of vorticity*. That fine weather in the heat low is maintained by subsidence resulting from upper-tropospheric convergence, and that there is lower-tropospheric convergence into the heat low, make it likely that there is net convergence in the tropospheric column and therefore that $A_\zeta < 0$ (Section 3.1.2).

(*ii*) *Diabatic influences*. Over a hot source, since $d\bar{H}/dt > 0$ and decreases outward, $\nabla^2 \, d\bar{H}/dt < 0$.

(*iii*) *Adiabatic changes*. Through the upper two-thirds of the troposphere, subsidence prevails. Therefore, $\omega > 0$, $(\Gamma_a - \Gamma) > 0$ and both probably decrease outward (clearest skies in heat-low center). In the lower third of the troposphere, this term varies diurnally (Fig. 3.4). During the day, $\omega < 0$ but $(\Gamma_a - \Gamma) \approx 0$. At night, ω and $\Gamma_a - \Gamma$ have the same sign as at the higher levels. Thus throughout the 24 hr, $\nabla^2 \langle \omega(\Gamma_a - \Gamma) \rangle_{av} < 0$.

(*iv*) *Thickness advection*. $A_T < 0$ and $\nabla^2 A_T > 0$ near the center of a heat low, which is weakened by cold-air advection.

Apparently, the diabatic and adiabatic terms overcome the effect of convergence and thickness advection to maintain the heat low.

The vorticity advection and adiabatic terms are interdependent, upper-tropospheric convergence leads to subsidence; the following evidence suggests that the latter may outweigh the former.

Raman and Ramanathan (1964) suggested that over the Indian and Burma peninsulas an upper-tropospheric easterly jet usually develops 100–200 km south of where the previous day's maximum rainfall occurred. The easterlies would speed up, apparently in response to a redistribution of heat of condensation. Possibly then, the easterlies would increase during periods of heavy rain, thus increasing downstream convergence and consequent subsidence over the heat low. Figure 3.5 indicates that surface pressure at Jacobabad, near the center of the West Pakistanian heat low may be inversely related to the intensity of monsoon *rains* over a strip of the subcontinent extending eastward to the south and east of the station, and Dixit and Jones (1965), while comparing a monsoon rain with a monsoon lull along the west coast of India, located by far the greatest middle- and upper-tropospheric temperature differences in the region above the heat low, with the rain situation 2–6C warmer than the lull situation. Hence, over the heat low, usually

$$A_\zeta - R/f\nabla^2 \log(p_0/p)\langle \omega(\Gamma_a - \Gamma) \rangle_{av} > 0.$$

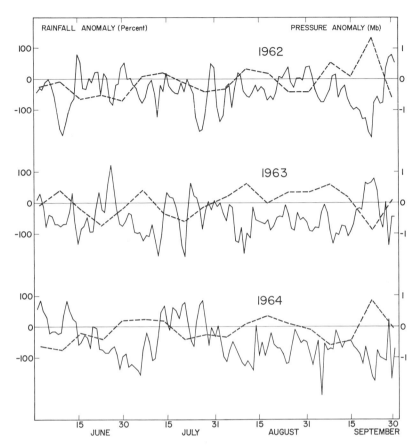

Fig. 3.5. Daily anomalies (mb) of mean sea-level pressures measured at 0800 local time at Jacobabad (28°18′N, 68°28′E; 56 m above MSL, full lines) and weekly percentage anomalies of rainfall for Indian subdivisions lying roughly between 18 and 27N (dashed lines) for the summers of 1962, 1963, and 1964.

In the heat low, during the great western India drought of 1899 (Ramage, 1969a), surface pressures were slightly above and surface temperatures were more than 1C below normal. Since no rain fell and the air was drier than normal, insolation must have been no less intense than usual. Apparently, a reduced cyclogenetic contribution from the monsoon *rains* overcame an enhanced contribution by sensible heat.

Similar developments occur over western Africa where disturbances resembling subtropical cyclones (Sections 3.5 and 5.7.1.2.; Figs. 5.13 and 5.14) move westward between 10 and 15N. Accompanying some of the disturbances is a nearly cloud-free surface depression traveling along the heat trough near 20N (Carlson, 1969b). Although surface pressure may fall and rise by as much as 4 mb with its passage, this depression vanishes at the coastline or stays over land as part of the heat trough.

3.3   Tropical Cyclones

These most destructive of synoptic-scale systems have also been the most intensively studied tropical meteorological phenomena (Palmén and Newton, 1969). They may develop and intensify to hurricane strength over the Pacific and Indian ocean parts of the monsoon region and dissipate over or near the bordering continents (Fig. 3.6).

3.3.1   DEVELOPMENT

Palmén (1956) listed the following prerequisites for the development of an intense tropical cyclone:

1. "Sufficiently large sea or ocean areas with the temperature of the sea surface so high (above 26 to 27C) that an air mass lifted from the lowest layers of the atmosphere (with about the same temperature as the sea) and expanded adiabatically with condensation remains considerably warmer than the surrounding undisturbed atmosphere at least up to a level of about 12 km. [Some evidence suggests that the absolute value of the sea-surface temperature may not be critical, but rather that the *gradient* of sea-surface temperature should be small over distances of several hundreds of kilometers.]
2. "The value of the Coriolis parameter larger than a certain minimum value, thus excluding a belt of the width of about 5 to 8 deg lat on both sides of the equator.
3. "Weak vertical wind shear in the basic current, thus limiting formation to latitudes far equatorwards of the subtropical jet stream."

Riehl (1954) listed these additional requirements:

4. A preexisting low-level disturbance (areas of bad weather and relatively low pressure).
5. Upper-tropospheric outflow above the surface disturbance.

A fully developed tropical cyclone is a warm-cored, energy-exporting system which usually remains intense for many days over the ocean. Essential to this is an extremely large surface pressure gradient near the core of the cyclone, sometimes exceeding 3 mb km^{-1}. Anthes and Johnson (1968) pointed out,

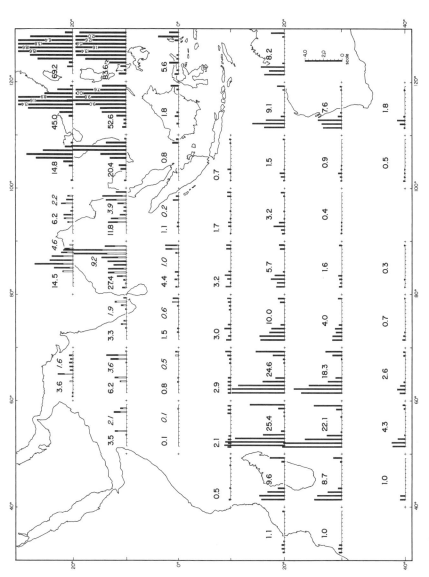

Fig. 3.6. Tropical cyclones. Average numbers of occurrences per 10 yr for each 10° square by months. The upright figures indicate average 10-yr totals. Over the Northern Indian Ocean the unblacked segments of the histograms and the sloping figures give corresponding information for severe tropical storms.

"As the air flows toward lower pressure, sensible heating serves to raise the equivalent potential temperature and is a necessary element in the maintenance of convection surrounding the eye. Ultimately, the latent-heat release occurring in the air with the higher equivalent potential temperature maintains the warm core, baroclinicity and the high efficiency factors, thus making the release of latent heat in this region more effective in the direct generation of available potential energy. Although the sensible heat generation is small, the role of sensible heating in maintaining the core of high equivalent potential temperature is an essential process in the energetics of hurricanes."

Simply put, an initial surface-pressure fall of 25–30 mb is needed before a self-sustaining hurricane can develop. Since divergence (pressure fall) in one atmospheric layer is almost invariably, and very rapidly, balanced by convergence (pressure rise) in another layer, a successful development theory must incorporate a lag between a large pressure fall at the surface and pressure rises above, sufficient to allow in-spiraling of air, massive rising in the central core, and the development, through subsidence, of a warm eye. When this sequence is followed, the pressure fall cannot be reversed and the hurricane is an accomplished fact.

Development may occur with almost explosive suddenness in less than an hour, or hesitatingly over several days. Detailed observations have yet to confirm whether one or a variable combination of the following might trigger a hurricane. In the upper troposphere, the cause may be dynamic instability (Sawyer, 1947) or energy dispersion (Ramage, 1959). Perhaps heat release in deep extensive cumulus convection could play a dominant part (see, for example, Yanai 1964). However, in at least one well-observed hurricane development (Hawkins and Rubsam, 1968) radar did not detect precipitation echoes above 10 km. Development generally seems to occur beneath upper-tropospheric flow from an equatorial quarter (Ramage, 1959; Colon and Nightingale, 1963).

As the vertical plane circulation of a monsoon system or a transequatorial Hadley cell intensifies, associated surface winds equatorward of a near-equatorial trough freshen and cause a local increase of cyclonic vorticity in the trough (Section 5.8.7.1). Occasionally a hurricane may develop there, perhaps as a consequence of these circulation changes whose causes are usually obscure.

Many recent attempts to model development numerically (Kasahara, 1961; Charney and Eliassen, 1964; Rosenthal, 1964; Kuo, 1965; Ooyama, 1969) have met with limited success. The rapid and essential initial pressure fall has not yet been simulated, nor can the numerical models be used to explain why development is so rare.

3.3.2 THE MATURE TROPICAL CYCLONE

Miller (1967) provided an excellent description of a mature hurricane, demonstrating that because of its warm-core character the intense cyclonic circulation diminishes with height, reversing to become predominantly anti-cyclonic in the upper troposphere. Air flows into the center at low levels and rises in the wall cloud surrounding the eye. Compensating outflow occurs in the high troposphere as well as some subsident inflow to the eye. The consequent warming maintains sea-level pressure in the eye below 950 mb. When measurements in a hurricane are made over a period of hours, the distribution of meteorological variables quite closely resembles long-term means (Fig. 3.7). However, large fluctuations may occur in the space of an hour or so, particularly in the core region of the storm, and the wall cloud might even temporarily dissipate.

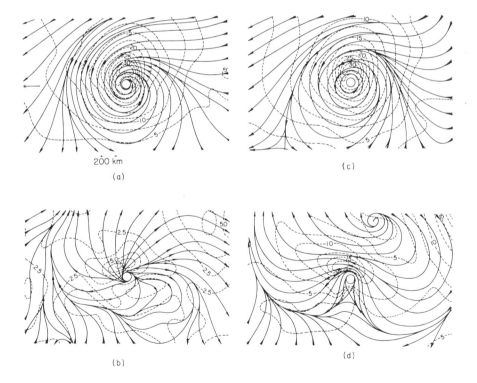

Fig. 3.7. Mean resultant hurricane winds (from Miller, 1958). Isotachs are labeled in meters per second. (a) wind field in 0–1 km layer (double shafted arrow denotes direction of storm center movement). (b) wind field in 3–6-km layer minus wind field in 0–1-km layer. (c) wind field in 3–6-km layer. (d) Wind field in 10.5–12-km layer.

3.3.2.1 *Accompanying Weather*. As is true of other cyclonic systems, the area of the hurricane circulation occupied by precipitating clouds seldom exceeds 15% (Gentry, 1964a). These clouds are concentrated in spiral bands, predominantly equatorward and westward of the center.

Ahead of tropical cyclones, just beyond the " bar " of the storm, surface divergence and upper-tropospheric convergence (Fig. 3.7) (Miller, 1958) clear the skies of convective clouds. This long-known effect (Hall, 1916) becomes sharply delimited as a storm reaches hurricane strength, when the cloud mass is drawn into a tight swirl around the center (Fig. 5.45), thus accounting for reports of temporarily *improving* weather in the path of an advancing, intensifying tropical storm. Occasionally, in a ring *surrounding* the storm clouds, skies are clearer than anywhere else in the enveloping tropical maritime air mass, indicating subsidence all around the circulation, which is then cut off from the band of clouds toward the equator and the west. Consequently, mean typhoon tracks almost never traverse regions of maximum average cloudiness (Plates I–IV).

3.3.3 MOVEMENT

Cyclones (including tropical cyclones) move toward falling pressures, that is toward upper-tropospheric divergence (Godske *et al.*, 1957, p. 555). In a climatological sense, this means in the general direction of the upper tropospheric current in the vicinity of the cyclone. Miller (1958) provided statistical confirmation, finding that hurricanes on an average moved with the mean flow in the 6–12.5-km layer in the ring of 200–450-km radius centered on the individual storm.

Demands for accurate forecasting have spurred vigorous research on tropical cyclone movement (Section 6.2). Improvement on the results obtained from simply extrapolating previous movement have proved difficult to achieve. At a major typhoon warning center, as recently as 1968, average errors in 24-hr forecasts of center movement using " persistence " were smaller than average errors resulting from application of sophisticated statistical and numerical techniques. Gentry (1964b) has written a useful survey paper.

For our purposes at this stage it suffices to point out that mean resultant 200-mb winds tend to parallel climatological tropical cyclone tracks.

The mechanism by which a hurricane converts heat energy into potential energy, and potential energy into kinetic energy, is not yet fully understood.

3.3.4 DISSIPATION

Hurricanes dissipate when the supply of sensible heat to the core is cut off, destroying the gradient of equivalent potential temperature at the surface

(Bergeron, 1954; Miller, 1964). This heat can be efficiently provided only by a uniformly warm ocean surface. The circulation weakens rapidly when it moves over land, or when a relatively cold or dry air mass penetrates the core.

The stress of cyclonic winds transports ocean-surface water outward from a tropical cyclone center (Section 2.2.1). Ensuing upwelling of a stratified ocean may lower the sea-surface temperature by as much as 5C (Leipper, 1967), but with a lag too great for the energetics of a *moving* cyclone to be affected. However, should the cyclone be stationary, it would be weakened as the gradient of surface θ_e decreased (Ooyama, 1969).

3.4 Monsoon Depressions

In summer the surface pressure trough or "monsoon trough" over the Indian subcontinent is oriented northwest–southwest lying to the south of and roughly paralleling the Himalayas (Fig. 5.42). The western end of the trough merges into the heat low over West Pakistan. The eastern end of the trough, over the northern Bay of Bengal is the birthplace of monsoon depressions (Fig. 3.19) which resemble both hurricanes and subtropical cyclones.

3.4.1 DESCRIPTION

Although monsoon depressions have never been probed by research aircraft, Indian meteorologists have studied numerous cases using surface and aerological observations made at stations in the paths of the circulations. From these studies it is possible to describe an *average mature monsoon depression* as it moves over land and slowly weakens.

The surface circulation, which sometimes covers more than 250,000 km² may possess gales over the sea, but over land winds seldom exceed fresh. Central pressures range from 2 to 10 mb below normal. Compared with typhoons, monsoon depressions appear weak, yet they often persist for a week or more over land while giving widespread and locally very heavy rain.

Early investigators claimed to detect significant air mass contrasts within depression circulations (Ramanathan and Ramakrishnan, 1932; Desai, 1951; Desai and Koteswaram, 1951) and related the distribution and intensity of rainfall to upslope winds at frontal surfaces. However, their observations were questioned by Mull and Rao (1949), and Rai Sircar (1956) whose opinion now prevails that air-mass discontinuities in monsoon depressions are as hard to detect as in typhoons (Koteswaram and George, 1958, 1960; Koteswaram and Bhaskara Rao, 1963; Ramaswamy, 1969).

Rai Sircar (1956) analyzed soundings made at Calcutta when three monsoon depressions ranging from weak to strong passed to the south. The local change of pressure at Calcutta is given by $\partial p/\partial t = \delta p/\delta t - \mathbf{U} \cdot \nabla p$, where $-\nabla p$, denotes the instantaneous pressure gradient, $\mathbf{U}$ the Calcutta-directed movement component of a passing storm, and $\delta p/\delta t$ differentiation following the pressure system. The term $\partial p/\partial t$ can only represent the instantaneous gradient of pressure $-\nabla p$ in a passing storm if $\delta p/\delta t = 0$ and $\mathbf{U}$ is constant. However, a storm approaching Calcutta over the northern Bay of Bengal usually intensifies and, as it passes away inland, is usually weakened. Thus in the lower troposphere both terms on the right-hand side would negatively contribute to $\partial p/\partial t$ as the storm approached and positively contribute as it receded. The effect would be to greatly enhance the apparent horizontal gradients of pressure—as indicated by the local changes at Calcutta. The intense monsoon depression analyzed by Rai Sircar (Fig. 3.8) was apparently

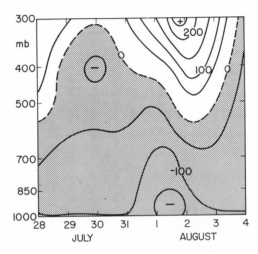

Fig. 3.8. Westward passage of a vigorous monsoon depression center 80 km to the south of Calcutta (22°39′N, 88°27′E) in July to August 1953 (Rai Sircar, 1956). Departures (in geopotential m) of Calcutta pressure heights from the mean tropical atmosphere (Riehl, 1954).

warm cored, with $\partial p/\partial t$ in the upper troposphere having the opposition sign from the lower troposphere, but with instantaneous pressure and temperature gradients much smaller than the local change would indicate. Evidence that the air below 3 km near the center was somewhat colder than the average for the storm circulation indicates that monsoon depressions might be most intense a kilometer or two above the surface (Desai, 1951) and resemble both hurricanes and subtropical cyclones (Fig. 3.1).

Pisharoty and Asnani (1957) found that heavy rains (often exceeding 10 cm in 24 hr) fall over a 400-km-wide strip to the left of monsoon depression tracks. The strip stretches from about 500 km ahead to about 500 km behind the center. Heavy rain does not fall within 60 km of the center (Bedekar and Banerjee, 1969).

3.4.2   Development

Before monsoon depressions develop at the head of the Bay of Bengal, lower-tropospheric southwesterlies freshen and weather deteriorates over peninsular India (Ramanathan and Ramakrishnan, 1932) and probably across the central Bay of Bengal (Section 5.8.7.1) while heavy rain sets in over the Andaman Islands and along the Tennasserim Coast (Desai, 1951). Cyclogenesis first appears at a height of 2–3 km and then extends to the surface (Desai, 1951). Other investigators confirm this sequence.

During the air mass vogue, cyclogenesis was thought to result from frontal wave development combined with cyclonic deflection of the fresh southwesterlies by the Arakan Mountains (Desai, 1951; Desai and Koteswaram, 1951).

Next, the middle-latitude development theories of Sutcliffe (1947), Riehl *et al.* (1952), and Petterssen (1956) (Section 3.1.2) were vigorously and ingeniously espoused by Koteswaram and George (1958, 1960) and Koteswaram and Bhaskara Rao (1963). They reproduced analyses to support their contention that a monsoon depression develops when simultaneously a surface isallobaric minimum or low-pressure wave moves from Burma to the northern Bay of Bengal beneath a layer of positive vorticity advection to the west of either a westward-moving trough or a jet maximum in the upper-tropospheric *easterlies*.

Of four or five low-pressure waves which move into the Bay of Bengal during an average summer monsoon month, two or three intensify into monsoon depressions (Koteswaram and Bhaskara Rao, 1963). Few of these pressure waves are remnants of Pacific typhoons which have moved westward from the South China Sea. In July and August for the years 1929–1938 (Chin, 1958; Ananthakrishnan, 1964), of 13 typhoons and tropical storms which crossed the Indochina coast, only 7 could conceivably have redeveloped into monsoon depressions. In the same months, 37 monsoon depressions developed over the northern Bay of Bengal.

At the International Meteorological Center, Bombay, well-documented twice-daily upper-tropospheric analyses extending eastward to 155E revealed both troughs and jet maxima in the 200-mb easterlies. But the troughs were usually temporary reflections of lower-level cyclones and moved erratically, while the maxima developed more or less *in situ* after heavy rains had begun 100–200 km to the north (Raman and Ramanathan, 1964).

After thousands of reconnaissance probes, the reason why tropical cyclones develop is still in dispute. Our knowledge of monsoon depressions is even more meager, although detailed observations may confirm the development theories of Koteswaram and his co-workers. About all that can be specified are favorable preconditions; the trigger remains undiscovered. First, the monsoon trough must lie some distance south of the Himalayas, generally along the Ganges Valley, and must protrude over the northern end of the Bay of Bengal; second, the vertical circulation over the southern Bay—convergent west–southwesterlies overlain by divergent east–northeasterlies must accelerate; third, the resultant increase in low-level cyclonic vorticity south of the monsoon trough is concentrated into a vortex in the trough, leading to massive lifting, release of condensation heat, and circulation intensification.

Monsoon depressions do not develop into tropical cyclones for two reasons: *first*, with the monsoon trough confined to the head of the Bay of Bengal, the sea cannot effectively supply the massive amount of sensible heat essential to a tropical cyclone; *second*, the presence of strong vertical wind shear (≈ 15 m sec^{-1} between 850 and 200 mb) (Section 3.3.1). This compares with shears over the central Bay of Bengal of 7 m sec^{-1} in April and 4 m sec^{-1} in November, and with 4 m sec^{-1} near Guam in July.

Monsoon depression development is part of the massive rhythm of summer rains over southern and Southeast Asia, which is discussed in Chapter 5.

3.4.3   MOVEMENT AND TRANSFORMATION OF MONSOON DEPRESSIONS

Monsoon depressions appear to move as tropical cyclones do—roughly west–northwestward along the tropospheric mean isotherms or parallel to the flow at 10–12 km (Ananthakrishnan and Bhatia, 1960). They generally merge into the monsoon trough (Section 5.8.7.5) over Rajasthan. However, about once in ten years a still-active depression recurves toward the north or northeast, giving heavy rain over Kashmir. Recurvature, occurring east of a large amplitude trough in the polar westerlies, accompanies a " break " in the monsoon, when rains shift north to the sub-Himalayas (Section 5.8.7.5).

Monsoon depressions, with some of the characteristics of subtropical cyclones (Section 3.5), have much smaller surface pressure gradients than hurricanes. Not only are they less energetic but also they do not depend significantly on addition of sensible heat to inflowing surface air to maintain the circulation (Section 3.3.1). As a depression moves over the flooded land, it experiences slight frictional weakening and weakens still more if moisture intake, and consequently condensation energy, is reduced.

Monsoon depressions dissipate only slowly. By the time one reaches Rajasthan, the moisture it has drawn westward may have aided subtropical cyclone development in the Bombay region (Section 3.5.4.2). The new system, tapping a fresh moisture source over the eastern Arabian Sea, and the old monsoon depression, together intensify the *rains*.

A synoptic example of a monsoon depression would not add much to what has already been said. Centers of well-developed depressions almost never pass close to rawinsonde stations (Fig. 3.8), they have never been probed by research aircraft, and are so enveloped in a chaotic mass of clouds that weather satellite pictures fail to reveal significant details (Fig. 5.43).

3.5 Subtropical Cyclones

Palmén (1949) described the sequence of events leading to the "cutting off" of a cold upper-level cyclone equatorward of the main polar westerly stream. Other papers by his group (Palmén and Nagler, 1949; Hsieh, 1949) contain detailed three-dimensional case history analyses of cutoff lows.

3.5.1 SUBTROPICAL CYCLONES OVER THE EASTERN NORTH PACIFIC

During winter, upper-level cutoff lows frequently develop over the eastern Pacific north of Hawaii and many are associated with surface cyclones. Known in Hawaii as "kona" storms, they occasionally cause widespread flooding in the island chain. Simpson (1952) called them "subtropical cyclones" and confirmed that they develop along the lines of Palmén's models, being always preceded by the injection of cold air aloft through the agency of large-amplitude troughs in the polar westerlies.

How can cold upper cyclones persist, intensify, and occasionally extend to the surface after being cut off from fresh supplies of cold air, when condensation and precipitation should inevitably weaken the thermal and pressure gradients of the system? This question, which Palmén's collaborators tried to answer, is more pertinent to subtropical cyclones, which last for days and even weeks without weakening and generally extend to the surface.

Ramage (1962) studied two subtropical cyclones in the vicinity of the Hawaiian Islands and concluded that they were in fact direct, energy-creating systems. A schematic cross section of a mature, steady-state subtropical cyclone, based on this small sample, is shown in Fig. 3.9. The middle troposphere between 400 and 600 mb is the layer of largest pressure gradients, strongest winds, and greatest convergence, all increasing from the periphery inward. The model is in a steady state, and so convergence leads to upward motion above the layer of maximum convergence, to downward motion beneath it, and to compensating divergence in other layers.

3.5.1.1 *Upward Branch.* Beyond about 500 km from the center, the upward motion is weak, resulting in only scattered upper clouds and dry adiabatic

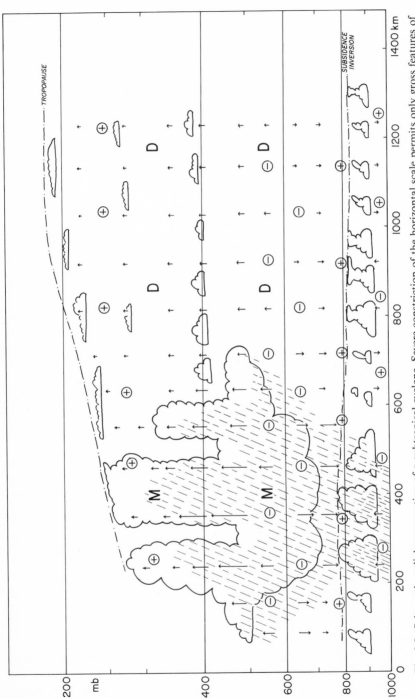

Fig. 3.9. Schematic radial cross section of a subtropical cyclone. Severe constriction of the horizontal scale permits only gross features of the vertical motion and cloud systems to be shown. Divergence is indicated by plus signs, convergence by minus signs. Regions of vertically moving air undergoing dry adiabatic temperature changes are denoted by D and regions undergoing moist adiabatic temperature changes by M.

cooling which further inhibits the upward motion. Closer to the center, vigorous upward motion leads to condensation and the development of deep precipitating clouds. Since the cooling is moist adiabatic, the circulation is direct and energy exporting. The upper branch, relatively warm cored, might be said to resemble a weak tropical cyclone circulation.

3.5.1.2 *Downward Branch.* Beyond about 500 km from the center, gentle downward motion accompanied by dry adiabatic warming extends to the subsidence inversion. As in trade wind regimes, some of the subsiding air probably leaks through the inversion. Beneath the upper rain clouds, the air descends more rapidly, is cooled by evaporation, and may become saturated. Since it then warms at a rate approaching the moist adiabatic lapse rate, it is denser, level for level, than the surrounding, dry, adiabatically warming air. Thus the downward branch is also a direct circulation.

3.5.1.3 *Surface Layers.* Beneath the subsidence inversion, a regime which bears some resemblance to the trade winds prevails. Local areas of wind maxima, possibly associated with surface heating, and corresponding zones of divergence and convergence are found.

3.5.1.4 *Eye.* From scanty evidence, the eye appears to be rather large, possibly more than 200 km in diameter, with generally scattered, haphazardly distributed clouds. If the hurricane analogy is valid, the eye in the upper troposphere should be relatively warm and cloud free. In the lower troposphere, the eye may slope with height. Various combinations of eye slope and eye diameter could result in widely varying eye clouds, as viewed from a weather satellite.

A subtropical cyclone must be distinguished from a large-amplitude trough in the polar westerlies associated with a surface low (in the latter systems, bad weather is generally concentrated east of the trough axis). Although a subtropical cyclone usually develops from such a trough and finally is absorbed by another, its field of motion, clouds, and weather are very differently and much more symmetrically disposed.

3.5.2 Subtropical Cyclones over the North Indian Ocean

In summer 1963, aircraft of the U.S. Weather Bureau Research Flight Facility (RFF) were deployed to the Indian Ocean to support the meteorological program of the International Indian Ocean Expedition (IIOE).

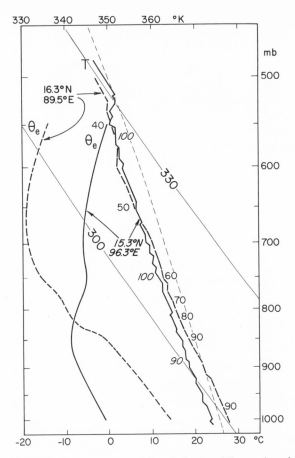

Fig. 3.10. Distribution of temperature (*T*), relative humidity, and equivalent potential temperature (*θ*ₑ) determined from RFF drop soundings on 1 June 1963, made west of a monsoon rain area (dashed lines) and within the rain area (full lines).

Reconnaissances across the central Bay of Bengal on 1 and 2 June revealed the existence of an extensive disturbance possessing many of the characteristics of a subtropical cyclone (Ramage, 1964c). A vigorous cyclonic circulation at 500 mb lay between surface westerlies and upper-tropospheric easterlies. The decrease in circulation intensity upward and downward from mid-troposphere was consistent with the temperature distribution: drop soundings made from the aircraft revealed that in the storm-rain area the upper troposphere was warmer and the lower troposphere colder than on the storm periphery (Fig. 3.10), where scattered cumulonimbus (*frontispiece*) reflected greater instability.

Below 800 mb, equivalent potential temperatures $(\theta_e)^*$ in the rain area were lower than on the periphery, and above 800 mb, higher. In the known absence of frontal systems, this difference can be accounted for only in terms of relative vertical motion (Riehl and Pearce, 1968). Compared with the periphery, lower-tropospheric sinking, brought about by mid-tropospheric convergence and evaporational cooling of the air, occurred in the rain area. The fact that θ_e was almost constant throughout the rain-area sounding suggests that the subtropical cyclone was efficiently transporting energy from the lower to upper troposphere (see Riehl and Malkus, 1961).

In the following month, a subtropical cyclone, which developed over the northeastern Arabian Sea, was intensively probed by the RFF and by the Woods Hole Oceanographic Institution's research aircraft. These unprecedentedly detailed observations were analyzed by Miller and Keshavamurthy (1968). Their model of a steady-state subtropical cyclone was derived by compositing the relatively unchanging fields of motion, pressure, temperature, clouds, and rain observed in and around the cyclone from 2 through 10 July 1963. In the following sections I quote liberally from this monograph.

3.5.3 THE MILLER–KESHAVAMURTHY SUBTROPICAL CYCLONE

3.5.3.1 *Distribution of Wind, Pressure, Temperature, and Moisture.* The composited cyclone (Figs. 3.11–3.14), at 700, 600, and 500 mb had centers in the *kinematic analyses* near 20.8N, 73E; 20N, 72.3E; and 18.8N, 72 E, respectively. The southward slope of the axis through the composite centers was not due to the method used in compositing, but rather is characteristic of mid-tropospheric cyclones over the west coast of India. The slope was related to the thermal distribution within the system and the effects of a heated land adjacent to a relatively cold sea.

In the mid-troposphere the composite circulation around the cyclone consisted of inner and outer rings of speed maxima with central values as high as 20 m sec^{-1} at 600 mb. In the outer ring the wind speeds were greater at 600 than at 500 mb, especially to the west and south. The trough in the wind field extended from the cyclone center to 65E, in good agreement with

---

* Total static energy of an atmospheric layer, $gz + c_p T + Lq$ (potential energy + enthalpy + latent energy of water vapor) closely approximates the total energy of the layer since the contribution of kinetic energy is small. The energy is nearly proportional to the equivalent potential temperature (θ_e), which can readily be determined from the plot of an aerological sounding (Byers, 1959), or more closely to the virtual equivalent potential temperature (θ_{ve}). Both θ_e and θ_{ve} are conservative for all dry- and saturation-adiabatic processes.

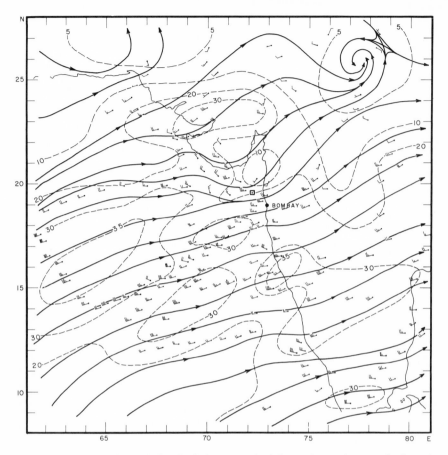

Fig. 3.11. Kinematic analysis of winds composited from observations made from 2 through 10 July 1963 for the low level (500–900 m). Each long barb in the wind plots represents 5 m sec^{-1}. Isotachs are labeled in knots (Miller and Keshavamurthy, 1968).

the composite pressure-height field. The circulation of the mid-tropospheric cyclone was delimited by anticyclonic flow on the periphery.

In the composited *pressure-height* field a definite pressure trough extended southwest from the center as far as 65E. At no level did pressure-height differences exceed 100 m between the cyclone center and the outer limits of the circulation. There was good agreement among the pressure-height, thermal, and wind fields within 400 km of the cyclone center.

A comparison of the streamline analyses with the contour analyses shows that a nonsteady state existed within 250 km of the cyclone center. The streamlines spiraling into the center of the lower pressure cut the contours at a sharp angle, especially in the southern half of the cyclone between 700

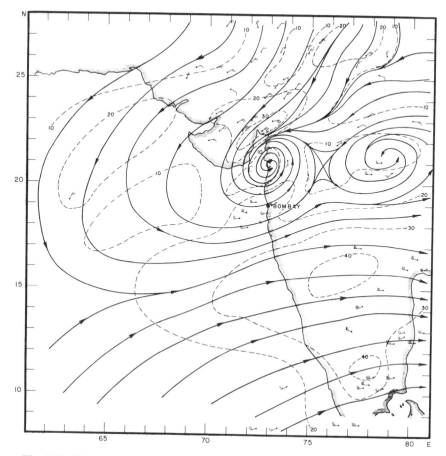

Fig. 3.12. Kinematic analysis of winds (same data and notation as in Fig. 3.11) for the 700-mb level (Miller and Keshavamurthy, 1968).

and 500 mb. This feature, which is observed in tropical cyclones, may be accounted for by the fact that the thermal field, within the most intense region of the cyclone, was in a constant state of flux due to changing intensities of rainfall, horizontal wind speeds, and vertical motions. In the outer periphery of the cyclone the general alignment of the streamlines and contours suggests better balance between the pressure force and the winds.

In the composite temperature fields (not shown), the center of the cyclone at 700 mb was colder than its environment; at 600 mb, it was neither warmer nor colder; and at 500 mb it was warmer. The system resembled the subtropical cyclone model described in Section 3.5.1, and lapse rates were less within the circulation than on the periphery (Fig. 3.10).

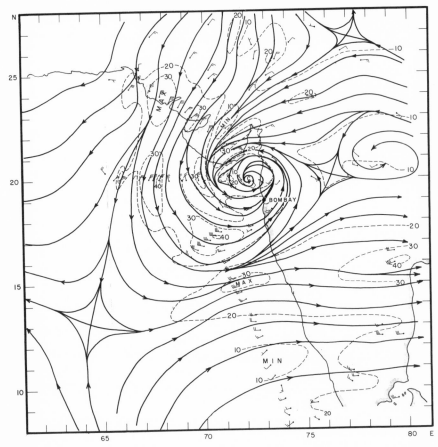

Fig. 3.13. Kinematic analysis of winds (same data and notation as in Fig. 3.11) for the 600-mb level (Miller and Keshavamurthy, 1968).

Comparison with the long-term mean temperatures reveals that the composite cyclone was about 4C colder than normal at 700 mb, 2C colder than normal at 600 mb, and near normal at 500 mb. On the periphery of the circulation temperatures were about normal.

High relative *humidities* were observed around the cyclone center, reflecting transport of moisture westward from a depression over the Bay of Bengal, and also from the south where moist air was being lifted off the Arabian Sea.

3.5.3.2 *Distribution of Divergence, Vorticity, and Vertical Motion.* From the composite streamline analyses, divergence and relative vorticity were computed and vertical cross sections constructed using the cyclone center

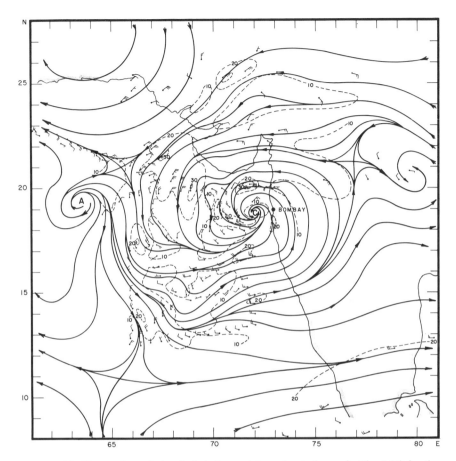

Fig. 3.14. Kinematic analysis of winds (same data and notation as in Fig. 3.11) for the 500-mb level (Miller and Keshavamurthy, 1968).

at 500 mb as the origin for east–west and north–south axes. From the cross sections of divergence vertical motions were computed.

(*i*) *Divergence.* Near the center of the cyclone between 600 and 500 mb, convergence exceeded 6×10^{-5} sec^{-1}. Above the level of nondivergence near 300 mb, divergence was as large as the convergence in the midtroposphere.

The *meridional cross section* (Fig. 3.15, top) shows that maximum convergence in the midtroposphere extended from 17 to 23 N. The divergence at 15N was associated with westerlies of 15–20 m sec^{-1} between 700 and 600 mb over the North Malabar coast. North of 25N, convergence below 700

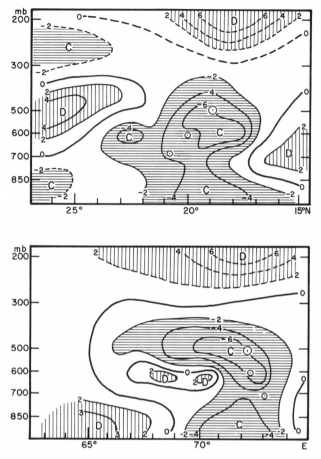

Fig. 3.15. Meridional cross section (top), and latitudinal cross section (bottom), of divergence (in units of 10^{-5} sec $^{-1}$) through the 500-mb composite center (18.8N, 72E) of the July 1963 subtropical cyclone. Above 300 mb, the dashed lines represent divergence computed from 2 July data only. The circled dots represent composite centers of the cyclone. The letters C and D denote centers of convergence and divergence, respectively (Miller and Keshavamurthy, 1968).

mb, middle-tropospheric divergence, and upper-tropospheric convergence were associated with the heat low over the subcontinent. There was low-level convergence at all latitudes, with a maximum in the surface layer below the cyclone.

The *latitudinal cross section* (Fig. 3.15, bottom) shows that, west of the center, convergence between 600 and 300 mb overlay divergence between 700 and 600 mb where stronger winds were found on the western edge of the cyclone system. Marked divergence west of 68E below 850 mb was associated with strong monsoon westerlies over the central Arabian Sea.

(*ii*) *Vorticity.* Cyclonic vorticity was greatest near the center of the cyclone, exceeding 16×10^{-5} sec^{-1} in the middle troposphere (Fig. 3.16). There was a westward and southward tilt, with height, of the axis of maximum vorticity between 700 and 500 mb, in agreement with the other composite fields.

Vorticity was anticyclonic in the upper troposphere.

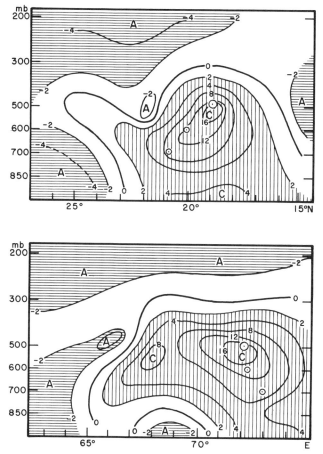

Fig. 3.16. Meridional cross section (top), and latitudinal cross section (bottom), of relative vorticity (in units of 10^{-5} sec^{-1}, through the 500-mb composite center (18.8N, 72E) of the July 1963 subtropical cyclone. Values above 300 mb are based on 2 July data only. The circled dots represent composite centers of the cyclone. The letters C and A denote centers of cyclonic and anticyclonic vorticity, respectively (Miller and Keshava-murthy, 1968).

(iii) Vertical motion (Fig. 3.17). In the composite cyclone, upward motions extended from the surface to above 200 mb and from the center outward to approximately 400 km. Upward motions in the cyclone circulation increased from 5 cm sec^{-1}, in the lowest levels and on the periphery, to 40 cm sec^{-1} above the center at 500 mb.

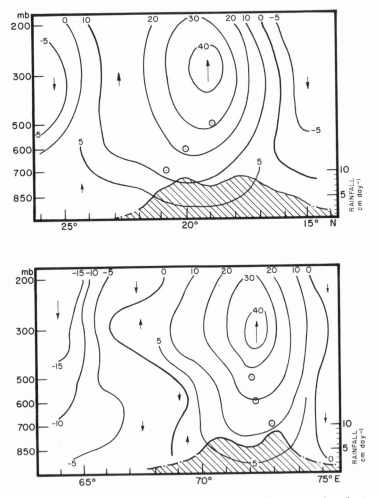

Fig. 3.17. Meridional cross section (top), and latitudinal cross section (bottom), of vertical motion (cm sec^{-1}) through the 500-mb composite center (18.8N, 72E) of the July 1963 subtropical cyclone, based on the composite divergence fields. The hatched area below the vertical motion profiles shows the meridional and latitudinal variation in rainfall along and across the Konkan coast based on a composite analysis of all daily rainfall amounts from 2 to 8 July at coastal stations (Miller and Keshavamurthy, 1968).

In the earlier subtropical cyclone model (Fig. 3.9), downward motion prevails between 600 and 800 mb and there is climatological evidence that surface divergence may occur beneath some Arabian Sea subtropical cyclones (Fig. 3.23). Miller and Keshavamurthy's model represents intense subtropical cyclones, which may even develop a small warm core surrounded by winds of gale force or stronger (Dixit and Jones, 1965; Simpson, 1952).

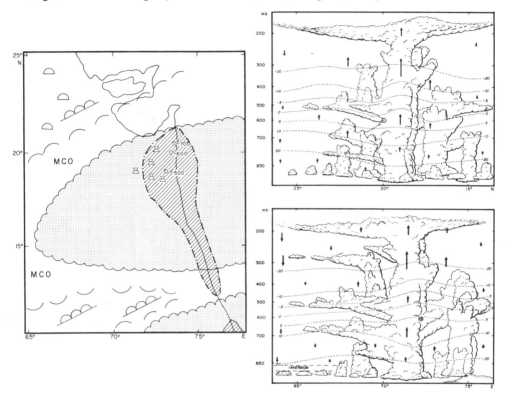

Fig. 3.18. Composite model showing the distribution of clouds, vertical motion, and temperature through the center of the July 1963 subtropical cyclone, with the longitudinal axis along 72E (top), and the latitudinal axis along 18.8N (bottom). The left-hand-side chart composites weather distribution charts constructed for individual days.

In the right-hand-side charts, clouds are shown schematically on a greatly exaggerated vertical scale. The dashed lines are isotherms taken from horizontal composite temperature charts. Solid arrows are vertical velocity vectors extracted from Fig. 3.17. The longest arrow is 40 cm sec^{-1} and the shortest, 5 cm sec^{-1}. The circled dot represents the 600-mb composite center of the cyclone. The hatched area in the bottom chart shows the extent of rainfall exceeding 4 cm day^{-1} associated with the subtropical cyclone. The region of greatest vertical cloud development is represented by cumulonimbus symbols. Stippled areas outline broken to overcast middle and high clouds; broken arcs show extent of broken to overcast low clouds and scattered high clouds (Miller and Keshavamurthy, 1968).

Beyond the limits of the cyclonic circulation, maximum downward motion occurred between 500 and 200 mb to the west of 65E, where air which had moved out of the top of the cyclone and downstream subsided. The marked subsidence inversions and extremely hazy conditions found on all research flights provide evidence of downward motion over the western Arabian Sea (Secton 5.8.7.4).

The levels at which maximum vertical velocities occurred were higher than in tropical cyclones.

3.5.3.3 *A Model of Clouds, Vertical Motion, and Temperature.* The composited fields of temperature, clouds, and precipitation, and horizontal and vertical motions combined with the cross sections of divergence, vorticity, and vertical motion (Figs. 3.15–3.17), describe a model of an Arabian Sea subtropical cyclone (Fig. 3.18).

3.5.4 DEVELOPMENT OF SUBTROPICAL CYCLONES

Synoptic analyses made at the International Meteorological Center (IMC), Bombay, during IIOE, demonstrated that most of the summer *rains* of western India and the northeastern Arabian Sea were associated with subtropical cyclone development, while long-term records show that surface cyclones are almost unknown over the Arabian Sea in summer (Fig. 3.19) (see also Anjaneyulu, 1969). Thus hints of a cause of subtropical cyclogenesis might first be sought in mean charts for midsummer (Figs. 2.3, 2.4, and 3.20; Plate III).

Average conditions in the surface layers and in the upper troposphere are unexceptional: westerlies lying south of the heat trough beneath easterlies lying south of the upper-tropospheric subtropical ridge. This distribution arises from a temperature gradient directed southward from the heat trough. Interest lies in the middle-troposphere (Fig. 3.20, top): two cyclonic cells in a trough located near 20N. The eastern cell nearly coincides with the region of most frequent monsoon depression development (Fig. 3.19) (Section 3.4.2). On the other hand, surface cyclones are rare beneath the western cell, but mid-tropospheric subtropical cyclones are common.

3.5.4.1 *Role of the Heat Low in Subtropical Cyclone Development.* The heat low is stationary; subtropical cyclones develop, intensify, and decay over a restricted area of the northeastern Arabian Sea south of the heat low. The tremendous insolational energy added to the atmosphere in the heat low results directly in no significant increase in the intensity of the low—surface pressure falls only 6 mb from mid-May to mid-July. The heat low is stationary

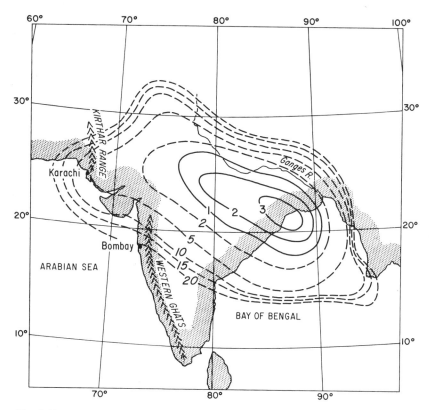

Fig. 3.19. Mean frequency of surface depression development during summer. Isopleths labeled in times per season (full lines) and in years between occurrences (dashed lines) (From Ramage, 1968a).

and always develops *before* the subtropical cyclones which produce western India's monsoon rains. The subtropical cyclones develop, intensify, and decay close enough to react on the heat low. Therefore, a hypothesis was advanced (Ramage, 1966) that energy exported from the heat low "causes" subtropical cyclogenesis.

Because of its nature the heat low is more likely to exert a thermodynamic rather than a mechanical effect on its environment. Petterssen (1956) discussed large-scale thermodynamic influences on the behavior of cyclones and anticyclones. He began with the vorticity equation for a surface of constant potential temperature and, by assuming steady state and neglecting the contribution of vertical advection of the tangential vorticity vector, derived the equation

$$\nabla \cdot (\zeta_a \mathbf{V}) = \zeta_F - (\partial \zeta_a / \partial \theta)(d\theta / dt) \qquad (3.4)$$

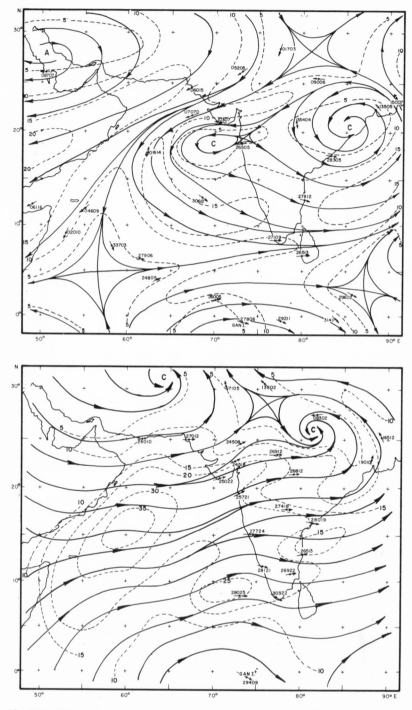

Fig. 3.20. Kinematic analyses of mean resultant winds for July at 900 m (bottom) and at 500 mb (top). Isotachs are labeled in knots (Miller and Keshavamurthy, 1968).

in which ζ_a is the component of the absolute vorticity normal to a surface of constant potential temperature.

Over Arabia and West Pakistan the heat low slowly intensifies from late May until mid-July then slowly weakens until it disappears after mid-September. Thus one might be justified in assuming steady-state conditions and in exploring the implications of Petterssen's equation.

The term on the left measures the intensity of the vorticity source, i.e., the export per unit time of vorticity through the boundaries of a unit area in a surface of constant potential temperature. The export depends on ζ_F, the intensity of the heat or cold source, $d\theta/dt$, and the sign of $\partial\zeta_a/\partial\theta$ which differs insignificantly from $\partial\zeta_a/\partial z$.

In only one part of the heat low are mean resultant tropospheric winds available for computing vorticity, the triangle enclosed by the rawinsonde stations of Peshawar, Jodhpur, and Karachi. Figure 3.21 depicts the mean

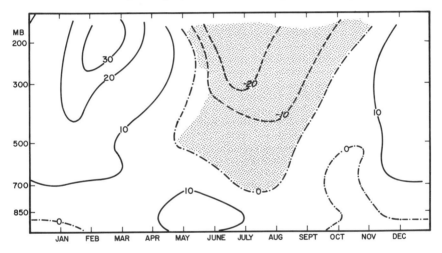

Fig. 3.21. Relative vorticity of the mean monthly resultant winds over the triangle Peshawar (34°01′N, 71°35′E), Karachi, (24°55′N, 67°08′E), and Jodhpur (26°18′N, 73°01′E) in units of 10^{-6} sec^{-1}.

distribution through the year and through the troposphere of the relative vorticity computed by the Bellamy (1949) method for the triangle, whose centroid is located at about 28N, 71E, on the axis of the heat low.

If ζ_F and $-(\partial\zeta_a/\partial\theta)(d\theta/dt)$ have the same sign, the sign of the vorticity export is known. During summer over the heat low, $d\theta/dt$ is always positive while, in the stippled area of the diagram, $\zeta_F > 0$ and $\partial\zeta_a/\partial\theta < 0$. Figure 3.21 shows that beginning in early June cyclonic vorticity is exported from the triangle in the middle and upper troposphere. If vorticity is similarly exported from Arabia, then cyclonic vorticity accumulates over the northern

Arabian Sea. As long as the air remains dry, development is unlikely, but by mid-June moist air reaches the northeastern Arabian Sea, usually from the south or east. Then buoyancy forces come into play, a subtropical cyclone develops, and the first burst of monsoon *rains* drenches the Bombay district. The subtropical cyclone through the agency of subsidence further intensifies the heat low which then exports additional cyclonic vorticity to the middle and upper troposphere over the northern Arabian Sea. It may be no accident that the rains of western India set in later (latitude for latitude) than the rains of eastern India and Burma, for the latter probably play a part in initiating the sharp increase in anticyclonic vorticity above the West Pakistanian heat low in late May.

Figure 3.21 shows that the depth of the layer of positive $\zeta_F - (\partial \zeta_a / \partial \theta)$ $(d\theta/dt)$ over the heat low decreases rapidly through September. At the same time, $d\theta/dt$ also decreases. By mid-September, in most years, positive vorticity is no longer exported to the northern Arabian Sea and no further subtropical cyclogenesis occurs there. One consequence of the model, indicated by crudely quantitative computations, is that the Arabian Sea subsidence inversion may so limit the supply of moisture as to make appreciably heavy monsoon *rains* over western India unlikely without significant incursions of deeply moist air from the Bay of Bengal across the peninsula. A burst of western India monsoon *rains* is usually preceded by increased depth of the moist layer to the east.

The foregoing discussion implies a symbiotic relationship between heat lows and subtropical cyclones: because cold upwelling coastal waters keep Somalia and Arabia dry and hot throughout the summer, because a complex of desiccating influences are at work over West Pakistan, and because topography concentrates midtropospheric cyclonic vorticity over the northern Arabian Sea, and because deep moist air can reach the region only from the south or east, the relationship is strongly developed over West Pakistan and the northeastern Arabian Sea.

3.5.4.2 *Development Sequence.* Miller and Keshavamurthy (1968) analyzed three subtropical cyclone sequences, including the case already described (Fig. 3.22). Although the sequences differed somewhat, in all cases middle-tropospheric development was preceded by cyclonic activity over eastern India, by above-normal, anticyclonic vorticity above the heat low, and by above-normal cyclonic vorticity between 18 and 21N. Then, as the subtropical cyclone intensified, anticyclonic vorticity above the heat low increased sharply, reflecting subsident warming and intensification of the surface circulation and of the southwest monsoon.

The possibility should not be ignored that general strengthening of one or more of the summer monsoon circulation modes (Section 5.8.7.1) could, by

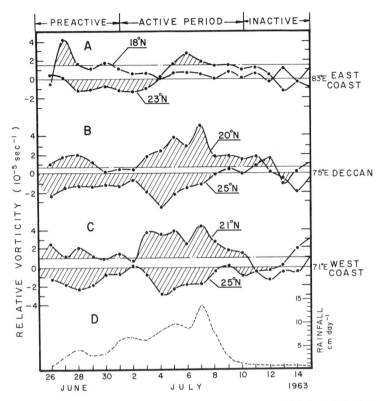

Fig. 3.22. Daily variations in relative vorticity from 26 June to 15 July 1963 in the 3–6-km layer between 18 and 25N along (A) 83E, (B) 75E, and (C) 71E. The horizontal broken lines represent the climatic mean layer vorticity for July at 20N; the climatic mean vorticity for July at 24N at each meridian coincides with the solid horizontal line of zero vorticity. Hatched areas represent periods when cyclonic vorticity exceeded the July mean vorticity (broken line) between 18 and 21N and when anticyclonic vorticity exceeded the July mean vorticity (solid zero line) between 23 and 25N. The active period extends over those days when rainfall was relatively heavy for prolonged intervals along the entire west coast. The daily rainfall amounts in (D) are averages based on 24-hr amounts from approximately 12 coastal stations between 15 and 20N (Miller and Keshavamurthy, 1968).

increasing cyclonic vorticity and moisture over the northeastern Arabian Sea, trigger subtropical cyclogenesis. Generally, however, the precedent is cyclonic activity over eastern India.

3.5.5 THE CIRCULATION ENVELOPE OF SUBTROPICAL CYCLONES

Figures 3.15–3.18 reveal some details of the links between a heat low and subtropical cyclone. Winds are predominantly zonal in the lower troposphere (westerly) and in the upper troposphere (easterly). Thus Fig. 3.17

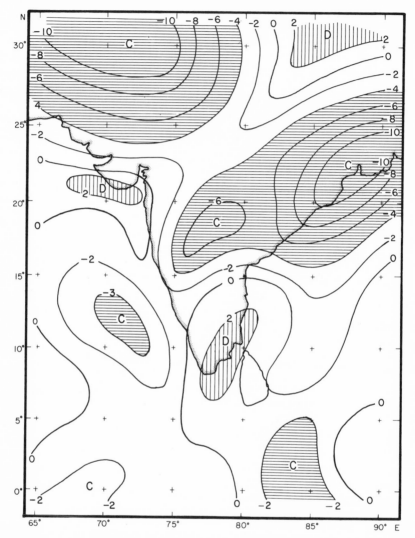

Fig. 3.23. July: divergence of mean resultant winds in the layer 0–900 m in units of 10^{-6} sec^{-1} (Miller and Keshavamurthy, 1968).

(bottom) shows a vertical circulation cell with sinking motion over the central Arabian Sea and rising motion in the subtropical cyclone. Upward speeds are greatest at 300 mb. In the lower troposphere weak downward motion may occur with less-intense cyclones.

The meridional circulation (Figs. 3.15, top and 3.17, top) is less readily interpreted since, except in the middle troposphere, distribution of divergence and vertical motion reflects gradients in the zonal component of the wind rather than in the much weaker meridional component. One might nevertheless postulate a two-celled vertical circulation. Climatologically this is supported by divergent mean resultant surface winds off Bombay (Fig. 3.23). Viewed from the west, two contrarotating helices interact at around 700 mb.

3.5.5.1 *Upper Helix.* Air which has spiraled vertically out of a depression over the northern Bay of Bengal or over the Ganges Valley flows westward in the upper troposphere, sinks above the Northwest Indian heat low, and then exporting cyclonic vorticity flows southwestward. Along with near-saturated middle-tropospheric air which has come directly from the monsoon depression, it is caught up in the vertical helix of the subtropical cyclone and brought once more to the upper troposphere whence it resumes its subsiding westward journey.

3.5.5.2 *Lower Helix.* Southwest monsoon air spirals upward in the North-west Indian heat low. Near the subsidence inversion it turns southward to flow beneath the subtropical cyclone. Part is incorporated in the upper helix, part sinks, and with some of the surface air which has come directly to the vicinity, swings northward to the heat low. Most of the vigorous southwesterly current sweeps onward beneath the subtropical cyclone to rise in the monsoon depression farther east.

Of course the circulation during a period of moderate monsoon *rains* is more involuted than this simple picture indicates, being as it is a complex of helices with not only zonal but also vertical axes.

3.5.6   SUBTROPICAL CYCLONES IN OTHER PARTS OF THE MONSOON AREA

Heat lows are a feature of the monsoon area, and rain-producing systems similar to subtropical cyclones have been observed south of the Saharan heat trough (Gilchrist, 1960; Carlson, 1969a) (Figs. 5.13 and 5.14) and north of the Australian heat trough (Falls, 1970). Subtropical cyclones may develop over the South China Sea and Indochina throughout the summer (Section 5.8.7.6). They have also been reported in early summer south of the thermal equator over the Bay of Bengal (Ramage, 1964c) and the Southeast Arabian Sea (Ramamurthi and Jambunathan, 1967). During the July to September dry season, a rare subtropical cyclone may move northwestward from the South Indian Ocean across East Africa (Gichuiya, 1970). (See Section 5.4.2.3).

In regions where the model depicted in Fig. 3.9 predominates, rainfall is unlikely to be positively correlated with low-level divergence (Dixit and Jones, 1965). Figure 3.24 (Rainey, 1963) shows divergence of 600-m mean resultant winds, and total rainfall for October 1954 over central Africa. In Fig. 3.25 rainfall is plotted against divergence for each rainfall station. Except

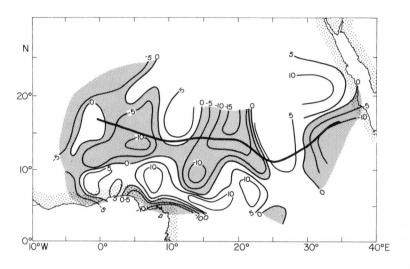

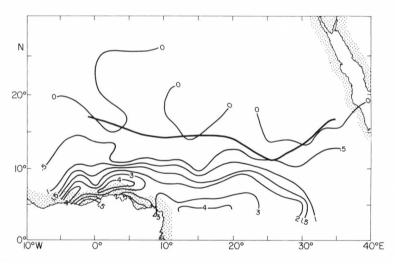

Fig. 3.24. October 1954: heavy line marks mean location of the heat trough line. (Top) divergence of the 600-m mean resultant wind field in units of 10^{-6} sec^{-1}. (Bottom) total rainfall in decimeters (Rainey, 1963).

at two coastal stations, Abidjan and Douala, heavy rain is likely to be associated with low-level *divergence* hundreds of kilometers south of the heat trough (median, 115 mm) and light rain is likely to be associated with low-level *convergence* in the vicinity of the heat trough (median, 17 mm).

An influx of dry air, *restricted* to the 850–700 mb layer might even favor subtropical cyclogenesis. Rapid evaporation of rain falling from above would

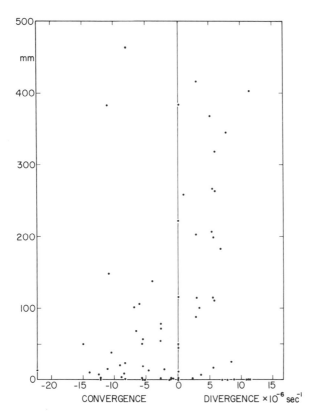

Fig. 3.25. Mean divergence of the 600-m wind field, and rainfall for October 1954 at individual stations in the area shown on Fig. 3.24.

cool this layer more than it would moist air and so would increase the vigor of the downward branch (Section 3.5.1.2). The dry air could be readily supplied from heat lows or from above the trade wind inversion. Could this be one reason why subtropical cyclones seem predominantly to develop near heat troughs or in regions dominated by the trade winds?

3.5.7   DECAY OF SUBTROPICAL CYCLONES

Unlike the winter subtropical cyclones of the northeastern Pacific (Ramage, 1962), those of the Arabian Sea and of other parts of the monsoon area do dissipate. Although a variety of causes might effect dissipation, the analogy of hurricane dissipation leads to a search for causes in the elimination of conditions which are apparently essential for development. Because feedback between a heat low and subtropical cyclone results in mutual intensification,

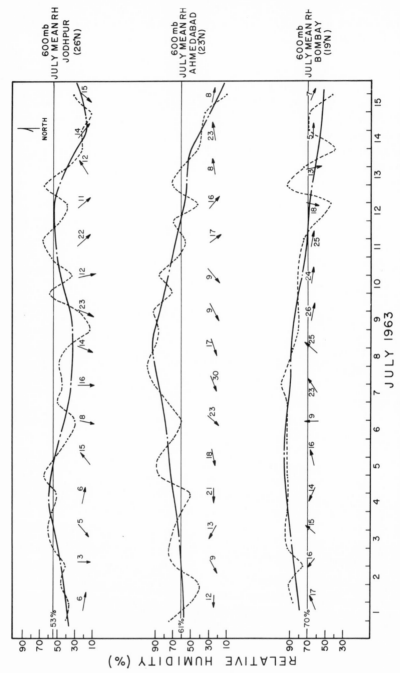

Fig. 3.26. Daily variations in the relative humidity (dashed lines) and winds at 600 mb for: Jodhpur (26°18'N, 73°01'E), Ahmedabad (23°04'N, 72°38'E), and Bombay (19°07'N, 72°51'E) for the period 1 through 15 July 1963. The horizontal full lines represent the July mean relative humidity at 600 mb above each station. The dot–dash lines show the running mean between successive days. Wind speeds in knots are given beside arrows showing wind directions at 600 mb (Miller and Keshavamurthy, 1968).

the possibilities are severely reduced. Miller and Keshavamurthy (1968)
ascribed dissipation to ventilation of the cyclone by drier air. A Bay of
Bengal monsoon depression which had facilitated westward advection of
moisture into the subtropical cyclone moved northeastward, moisture advection was cut off as weak northwesterlies were established in the middle troposphere over northern India. Figure 3.26 shows how relative humidities
responded at Jodhpur, Ahmedabad, and Bombay, and Fig. 3.22 depicts the
sharp decrease in circulation intensity and rainfall between 8 and 9 July. Dry-
air ventilation over the Indian subcontinent may stem from other causes.
Ramaswamy (1962) suggested that large-amplitude troughs in the mid-latitude westerlies might advect cold dry air south of the Himalayas, a formidable
barrier but not necessarily an impassable one (Section 5.8.7.5).

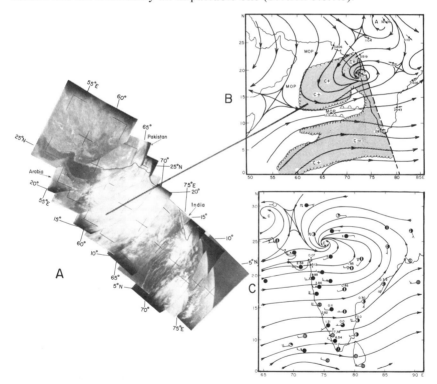

Fig. 3.27. (A) Clouds photographed from TIROS VII at 1040 GMT, 12 August 1964
(Orbit No. 3409). (B) Nephanalysis based on the photographs; 500 mb observed winds in
knots and streamlines for 1200 GMT. (C) 1500-m observed winds and streamlines, weather
and total cloud cover for 1200 GMT; rainfall amount in centimeters for the 24-hr period
ending 0300 GMT, 12 August entered above each station. The heavy dotted line running
between Charts (A) and (B) points out the extensive shield of middle and high clouds
associated with monsoon *rains* over the Arabian Sea (Miller and Keshavamurthy, 1968).

Whenever subtropical cyclones and heat lows adjoin, slight barely detectable shifts in the direction of mid-tropospheric inflow could rapidly inject very dry air into the subtropical cyclone.

Figures 3.27 and 3.28 (Miller and Keshavamurthy, 1968) show weakening of a subtropical cyclone. On 12 August 1964 thick nimbostratus surrounded the center at 500 mb and was giving heavy rain. On 13 August, neither the circulation at 500 mb nor the sounding at Bombay had changed appreciably. However, the weather satellite revealed predominantly cumuliform cloud. Disappearance of the nimbostratus indicated a decrease in the moist layer depth which was reflected in greatly diminished 24-hr rainfall measured on the morning of the 14th. The regime had changed from *rains* to *showers* (Section 4.1.2).

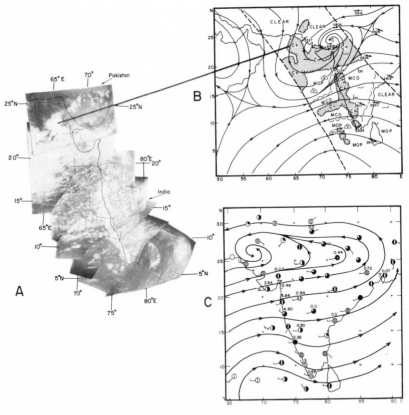

Fig. 3.28. Composite (A), nephanalysis (B), and wind chart (C) as in Fig. 3.27 but for 13 August 1964 (TIROS VII Orbit no. 3423 at 0950 GMT). The heavy dotted line running between charts (A) and (B) points out the shrinking shield of middle and high clouds associated with a weakening subtropical cyclone. To the south, convective cells replaced unbroken cloud (Miller and Keshavamurthy, 1968).

3.6   Weak Low-Pressure Circulations

Only a small proportion of low-pressure systems are intense enough to be classed as monsoon depressions, subtropical cyclones, or hurricanes. However, weak cyclonic circulations which appear in the same regions and in the same seasons as the strong circulations generally resemble them in a blurry sort of way, moving along similar tracks and at similar speeds (Ramage, 1951).

3.6.1   EASTERLY WAVES

Attention has been lavished on one such system—the *easterly wave*. First identified in the Caribbean Sea, the easterly wave as described by Riehl (1954) is a westward-moving disturbance of the tropical lower-tropospheric easterlies. The disturbance axis, usually oriented SSW–NNE, coincides with a pressure trough. As it passes, local winds shift from northeast to southeast. Since most easterly waves move more slowly than the environmental air, changes in vorticity along air parcel trajectories imply sinking motion and fine weather west of the axis and rising motion and poor weather east of the axis (Section 6.2.1.1). This distribution is commonly observed. According to Riehl, easterly waves are most intense in the middle troposphere where a cyclonic circulation often exists.

East–west-oriented lower-tropospheric troughs with easterlies poleward and westerlies equatorward of the axis are a feature of the monsoon area for almost all the year. Consequently, increased cyclonic vorticity is immediately manifested as a vortex and not as a wave. Very rarely, easterly waves have been observed in the monsoon area below 10° latitude in winter. Weather associated with weak vortices sometimes resembles easterly wave weather, although more generally it appears as a round or elliptically shaped cloud mass in weather satellite photographs.

3.7   Troughs in the Upper-Tropospheric Westerlies

The role of these troughs in middle-latitude weather processes and in surface cyclogenesis has been lucidly expounded by Sutcliffe (1947), Scherhag (1948), and Petterssen (1956). In similar fashion the troughs affect winter monsoon weather. Bjerknes (1951) emphasized that surface cyclogenesis may result either from dynamic instability associated with a frontal surface or from unstable growth of an upper wave trough. Both causes seem to operate in cases of intense cyclogenesis, or, as Petterssen said:

"Overtaking by an upper trough (with positive vorticity advection in advance of it) of a frontal system in the lower troposphere is one of the most reliable indications of cyclone development at sea level."

Except during winter off Southeast Asia, sharp fronts do not penetrate far into the monsoon region. With only one of Bjerknes' criteria likely to be satisfied, vigorous cyclogenesis east of an upper tropospheric westerly wave is unlikely until the surface disturbance has moved out of the tropics. However, even weak cyclogenesis significantly changes the weather.

3.8 Near-Equatorial and Monsoon Troughs

3.8.1 WEATHER DISTRIBUTION

Heat lows are fine-weather phenomena. Conversely, near-equatorial troughs are associated with unsettled weather. However, in a climatological sense, in the summer hemisphere they together comprise a continuous low-pressure belt (Figs. 2.1 and 2.3). The division between the two is found in the Northern Hemisphere summer over northwestern India, and in the Southern Hemisphere summer, some distance inland from the east coasts of Australia and southern Africa. In the summer hemisphere, the troughs lie close to the meteorological equator. In the winter hemisphere, a weak discontinuous trough lies a few degrees from the equator, close to a secondary meteorological equator. Sadler (1963) pointed out that even over the sea, there is less cloudiness at the trough axis than to north or south, a relationship also evident on mean charts (Plates I–IV).

In the heat lows, cloudless skies result from general subsidence in the latitudes of the subtropical anticyclones. But in the winter hemisphere or over the ocean in the summer hemisphere, the troughs lie equatorward of the ridges, separated from them by a cloudy zone.

Perhaps, even in the winter hemisphere, the troughs are partially under thermal control (Ramage, 1968a). Surface air converging into a trough ascends, and clouds, rain, and occasionally vigorous cyclonic circulations develop. These, by interrupting incoming solar radiation (Bunker and Chaffee, 1970) sharply reduce the surface heating needed to maintain the trough. On the edge of the cloudy zone, where sinking $[\nabla^2 \langle \omega (\Gamma_a - \Gamma) \rangle_{av} < 0]$ clears skies and leads to surface heating $[\nabla^2 (\partial \bar{H}/dt) < 0]$ a *new* trough develops. Bordering cloudy zones then shift correspondingly. This sequence could account for reports that the trough often appears to move discontinuously (Johnson, 1962; Thompson, 1965).

Over northern and northeastern India, where summers are rainy, the mean surface or monsoon trough, extending east–southeastward from the heat low over West Pakistan coincides with a relative cloudiness minimum (Sadler, 1969) (Fig. 2.3 and Plate III), a relative rainfall minimum, and a relative thunderstorm maximum (Fig. 5.42). The same forces that produce a relative cloudiness minimum at the near-equatorial trough over the ocean are undoubtedly at work here, anchored, and their climatological effect accentuated, by the Himalayas (Section 5.8.7.5).

Thus in the mean, maximum rains and troughs cannot quite coincide (c.f. Riehl, 1954), and the terms "near-equatorial trough" and "near-equatorial convergence" are not synonymous.

In a near-equatorial or monsoon trough weather varies sharply from one day to the next or at any one time along the trough. The trough is a favored place for tropical cyclone or monsoon depression development (Sections 3.3.1 and 3.4). When this occurs, weather deteriorates in the region of the circulation, but may frequently *improve* over that section of the trough toward which a vigorous tropical cyclone is moving (Section 3.3.2.1).

3.8.2   Theoretical and Numerical Studies

Careful observation and description, based especially on weather-satellite photographs, have drawn theoreticians' attention to near-equatorial troughs and the associated convergence lines (Charney, 1967; Pike, 1968; Krishnamurti, 1969). The tendency for cloud lines to persist along favored latitudes greatly intrigued them.

Charney and Pike dealt with the simplest case (sometimes occurring in the monsoon area during the transition seasons) of trade winds flowing toward equatorial low pressure.

Using linear analysis, Charney derived an expression for the maximum growth rate of a near-equatorial convergence:

$$\{[\tfrac{3}{2}\mu(1 + \gamma - \gamma_s) - 1]/[(1 - \mu(1 + \gamma - \gamma_s)]\} \{[C_D|V_{g_0}|(g/R\overline{T})]\} \qquad (3.5)$$

where $\overline{T}$ is the vertical mean of temperature and the lapse rates are measured in degrees Celsius per 100 dynamic meters.

$\gamma > \gamma_s$ in the maritime tropics. Pike (1968) pointed out, in discussing the left-hand term, that growth could occur only when $\tfrac{2}{3} < \mu(1 + \gamma - \gamma_s) < 1$, and be favored where surface temperature is highest. The contribution of the right-hand term (assuming a constant drag coefficient and a maximum of $R\overline{T}/g$ in the near-equatorial convergence region) depends on the magnitude of V_{g_0}. Pike concluded therefore that "growth would probably be most favored somewhere between the sea-surface temperature maximum and the geostrophic wind maximum."

Observation that the surface-pressure trough, which lies close to the sea-surface temperature maximum, does not coincide with the zone of maximum cloudiness (Section 3.8.1) lends credence to this interpretation.

Pike (1968) proceeded then to use nonlinear numerical techniques to compute vertical motion in a zonally invariant trade wind system. His primitive momentum equation model integrating at $7\frac{1}{2}$-min time steps was applied to different but constant surface-temperature distributions.

3.8.2.1 *Distributions Symmetric about the Equator.* With maximum surface temperature at the equator, two updrafts developed 6° to the north and south with sinking at the equator. After 24 hr, however, the updrafts moved together and from 30 hr onward remained over the equator. According to Pike, low-level frictional convergence prevailed at first, to be overcome by thermal forcing. A similar but less rapid sequence was computed for a uniform surface temperature field.

With temperature maxima at 8° latitude and a secondary minimum 2.5C lower at the equator, computations led to symmetrical updrafts which fluctuated between 6 and 12° latitude through 120 hr, and sinking at the equator.

3.8.2.2 *Distributions Asymmetric about the Equator.* Pike's computations for a fixed temperature field with a maximum at 10° latitude in the summer hemisphere produced an updraft fluctuating between 2 and 12° latitude in the *winter* hemisphere and a more variable updraft near the temperature maximum in the summer hemisphere.

3.8.2.3 *Comparison of Computations with Climatology.* Over the eastern central Pacific, where Pike's initial conditions are most closely approximated, cloudiness and rainfall are rather well correlated. On an average, in the transition months of April and October, Northern and Southern Hemisphere trades are about equal, although the surface-pressure trough lies slightly north of the equator. The sea-surface temperature distributions between 140 and 170W in the two months are quite different. In April, waters are warmest along 10S with equatorial gradients somewhat smaller than to north and south. In October, waters are equally warm along 10N and 10S, with a minimum along the equator. Figure 3.29 shows the meridional profiles of mean cloudiness for the two months.

In *April*, corresponding to Pike's asymmetric case, agreement between observation and model is rather encouraging—a sharp maximum in the colder hemisphere, a diffuse maximum in the summer hemisphere. That the equatorial minimum is rather better defined than the model indicates might be accounted for by the small temperature gradient.

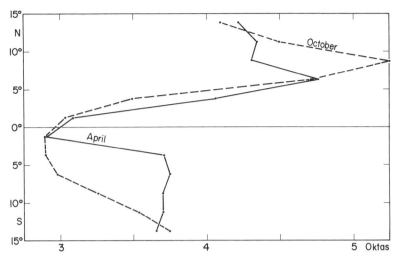

Fig. 3.29. Meridional profiles of average cloudiness between 140 and 170W for April and October, based on 3 yr of daily weather satellite pictures.

In *October*, corresponding to Pike's symmetrical, cool-equator case, the Northern Hemisphere cloudiness maximum and equatorial cloudiness minimum fit the model well. But where is the Southern Hemisphere maximum? Average cloudiness increases steadily from the equator to at least 30S. Sea-temperature and cloudiness profiles for March and May resemble April; those for August, September, and November resemble October.

Krishnamurti (1969) applied the primitive momentum equations to the situation over the western Pacific at 1200 GMT, 1 March 1965 and, using 10-min time steps, made 6-, 12-, 18-, and 24-hr forecasts of the winds, vertical motion, and convective cloudiness. His main purpose was to study the near-equatorial convergence zone north of the equator. Since the numerical prediction did not destroy the large-scale features (surely a negative criterion), Krishnamurti considered that the predicted fields would describe the large-scale dynamical structure of the convergence zone. The convergence zone remained in the forecasts and the 24-hr forecasts of the 1000- and 200-mb circulations looked reasonable. Observations have a way of disagreeing with forecasts and this was no exception. However, the results were not discouraging.

Pike (1970), by presuming the sea to react to the winds, developed a more realistic model in which equatorial cooling resulting from upwelling in turn affected the vertical motion of the overlying air.

Using numerical modeling, meteorologists are making the first rough diagnoses of near-equatorial troughs where the circulation is simple and the surface homogeneous. The technique will eventually be applied to the monsoon area, but for the time being results have limited applicability, particularly over rugged continents where insolationally controlled surface-temperature fluctuations must rapidly and significantly react on the circulation.

3.9   Quasi-Stationary Noncirculating Disturbances

Tropical cyclones and monsoon depressions associated with heavy extensive rains *move* much as middle-latitude cyclones move, although more slowly. Consequently most tropical meteorologists used to assume that all areas of low-latitude bad weather also moved.

Middle-latitude fronts lie in troughs and move and persist as air-mass boundaries. Some meteorologists, considering climatological evidence for a near-equatorial trough, thought that from day to day a front would be found in it, separating Northern Hemisphere air from Southern Hemisphere air and behaving similarly to fronts in higher latitudes.

For many years, the concept of moving, persistent synoptic entities usually formed the basis for tropical forecasts. Despite meager success, sparse observations allowed meteorologists to cling to the belief that advective changes far outweighed time changes in contributing to local weather changes. The many forecast errors were often excused on the ground of randomly occurring but intense convection, while a disturbance was said to "move in" when the rain started and to "move away" when the rain ended.

3.9.1.   Observational Evidence

Although Braak (1921–1929) reported that rain areas seldom progressed across Indonesia (Section 5.5.3), few meteorologists apparently heeded him.

During World War II, observations in the tropics were often sufficiently numerous to throw serious doubt on the comfortable notions of disturbance movement or of random convection. I vividly recall my first direct experience of a "jumping" or "disappearing" "intertropical front," and my embarrassment on another occasion when for 8 hr rain deluged the equatorial coral speck for which I was the resident forecaster.

Shortly after the war an excellent observing network was established over equatorial Africa and revealed, according to Johnson (1962),

" . . . lack of evidence for advection of the large-scale rain systems. . . .
They tend to develop *in situ* during the course of one or two days, and then persist for a period with minor changes, before declining . . . mobile perturbations such as equatorial or easterly waves contribute little to East African weather."

Johnson also showed that "large [rain] falls do not result simply from chance local instabilities, but are usually engendered during intense large-scale development." Earlier, Thompson (1957a) asserted that "the application of simple Intertropical Convergence Zone theories in forecasting practice in East Africa has failed lamentably."

The weather satellites have now confirmed that synoptic weather patterns over Indonesia and eastern Africa, far from being exceptional, are typical of very low latitudes. Bad weather usually develops, intensifies, and dissipates *in situ* sometimes in meso- or synoptic-scale systems of approximately circular shape, without observable circulations (Simpson *et al.*, 1967), sometimes in east–west lines thousands of kilometers long (Tschirhart, 1959). In fact, equatorward of 10° occasional squall lines, chiefly confined to western Africa, and a rare tropical cyclone are about the only *moving* disturbances recorded by the all-seeing orbiting cameras.

3.9.2 SEARCH FOR EXPLANATIONS

Quasi-stationary subtropical cyclones (Section 3.5) sufficiently distort the normal environment for their meteorological gradients to be measurable by conventional observation and reconnaissance equipment. Much remains unknown about them but nevertheless consistent descriptions have been made that are suitable for quantitative testing with diagnostic numerical models.

On the other hand, no such success has been attained with quasi-stationary disturbances in equatorial regions. Meteorological gradients, apart from rainfall gradients, associated with the disturbances are seldom large enough to be distinguishable from the gradients associated with undisturbed situations (Simpson *et al.*, 1967) while reasons for development and decay are obscure.

Work in East Africa (Johnson and Mörth, 1960) and in Southeast Asia (Ramage, 1968b) suggests that widespread but subtle changes in stability of the equatorial atmosphere result from changes in higher latitudes, particularly near the subtropical jet stream. An efficient draining away of surplus heat in the equatorial upper troposphere might sufficiently reduce the stability of the air column to initiate widespread convection. Conversely, an accumulation of heat of condensation in the equatorial upper troposphere might well stop or inhibit widespread precipitation.

3.9.3. EXAMPLE: 6 THROUGH 14 MARCH 1967

Following a surge of the winter monsoon, northerlies freshened suddenly (Section 5.8.2.3) over the South China Sea during the 5th and 6th. Bad

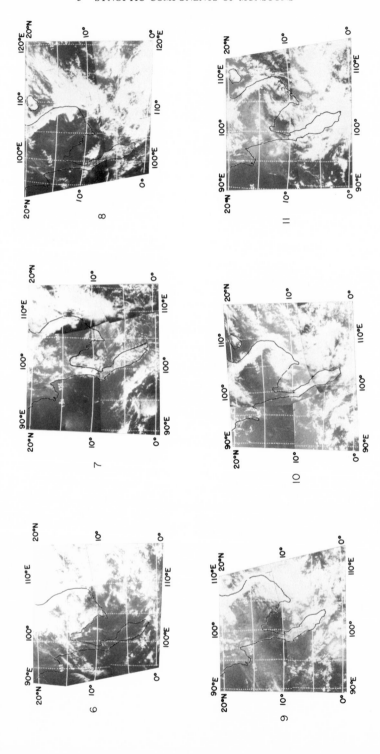

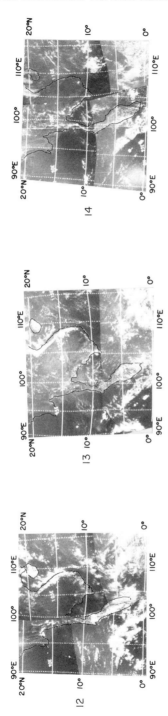

Fig. 3.30. Clouds photographed from ESSA 3 between 0500 and 0800 GMT, each day from 6 through 14 March 1967.

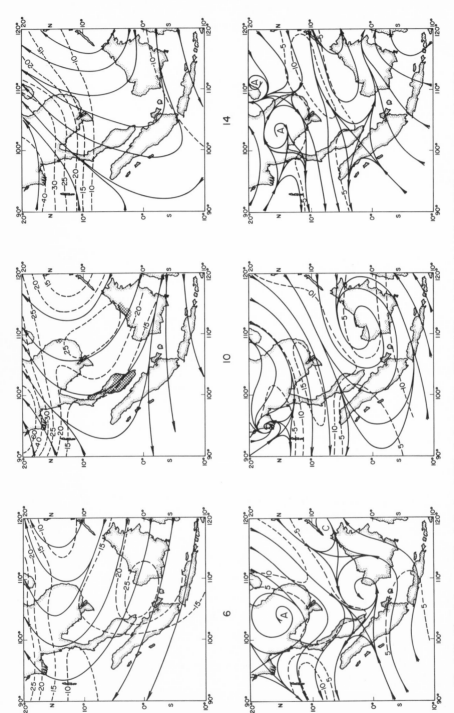

Fig. 3.31. Circulations at 200 mb (top) and 850 mb (bottom) for 0000 GMT on 6, 10, and 14 March 1967. Isotachs are labeled in meters per second. In the upper charts, crosshatching delineates regions in Burma, Thailand, South Vietnam, Malaya, and North Borneo with 24-hr rainfall totals exceeding 15 mm.

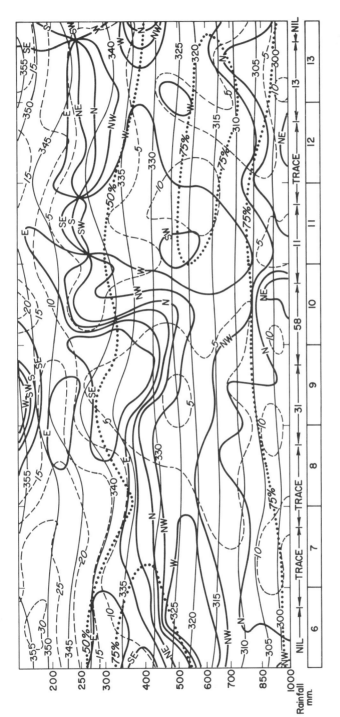

Fig. 3.32. Time cross section for Singapore (1°21'N, 103°54'E) 6–14 March 1967. Isogons are shown thick and full, isotachs (m sec⁻¹) dashed, isentropes (°K) thin and full, and isohumes (percent) dotted.

weather which had developed on the 4th and 5th over North Borneo spread to Malaya on the 9th. By the 13th conditions had improved. The sequence is illustrated by daily weather satellite pictures (Fig. 3.30) and by analyses of the lower- and upper-tropospheric circulations on the 6th, 10th, and 14th March (Fig. 3.31). Continuity is provided by a time cross section for Singapore (Fig. 3.32).

3.9.3.1 *6 through 8 March*.

Heavy falls at North Borneo stations on the 4th and 5th coincident with passage of a cold surge across South China conformed to the pattern discussed in Section 5.8.2.3 and shown in Fig. 5.20. Between the 6th and 8th, although surface winds exceeded $10 \, \text{m sec}^{-1}$, the monsoon flow was not particularly deep. Weather temporarily improved over North Borneo on the 6th. In the upper troposphere strong easterlies at Singapore indicated upstream divergence balancing lower-tropospheric convergence into the near-equatorial trough.

Changes to the north and south of Malaya during the 8th might be said to account for the expansion and intensification of the bad weather area. *In the north* 200-mb southwesterlies at Bangkok temporarily increased, reaching $50 \, \text{m sec}^{-1}$ at 0000 GMT on the 9th, while above 200 mb, westerlies replaced easterlies and temperatures fell as far south as Singapore. Presumably a trough in the upper-tropospheric westerlies, moving slowly eastward across the Bay of Bengal, had intensified. In the lower troposphere the monsoon freshened and deepened.

In the south, the near-equatorial trough remained near 10S with most cloud concentrated in westerly flow north of the trough axis. In the upper troposphere east–northeast winds doubled in strength. Information is too scanty to determine whether a Southern Hemisphere cold outbreak coincided with this change.

3.9.3.2 *9 through 11 March*.

Malaya and Borneo now lay in a region of upward motion common to a Northern Hemisphere Hadley cell and a more complex Southern Hemisphere vertical circulation, a situation not dissimilar to the one accompanying torrential rain in Kerala in October 1964 (Section 4.2.2.2). Kota Bharu (6°10′N, 102°17′E) remained in the northern Hadley cell with persistent upper-tropospheric southeasterlies; at Singapore, wind directions fluctuated about east, presumably in response to meridional fluctuations in the boundary zone between the cells.

Over Malaya heaviest rain fell on the 10th between 2 and 5N, with the greatest 24-hr amount (316 mm) recorded at Kuantan (3°46′N, 103°12′E). On the 11th rain was still considerable. The region of heaviest falls had shifted about 200 km north—perhaps corresponding to a backing in the upper tropospheric winds over Singapore.

Although moisture content over Singapore changed very little from the 6th to the 14th, upper-tropospheric stability decreased significantly during the 9th when the rain was developing and was not restored until the 14th. This means that the circulation cells by efficiently draining away the heat generated by condensation maintained a lapse rate favoring further vertical motion and rain (Sections 4.2.2.2 and 7.2.1).

The extensive amorphous cloud masses in the satellite photos typified a *rains* regime (Section 4.1.1) (Fig. 3.27), while the importance of synoptic-scale rising motion is additionally emphasized by the fact that rain to leeward of the main ranges was almost as heavy as to windward.

3.9.3.3 *12 through 14 March.* The anticyclone over China began moving eastward on the 11th, the Northern Hemisphere Hadley cell weakened, and upper-tropospheric winds south of the equator veered toward south. Rain diminished; predominantly cellular clouds in the satellite pictures reflected transformation to a *showers* regime (Section 4.1.1.).

In the course of a week, in which no disturbance entered or left the region and the near-equatorial trough remained stationary, a mass of cloud formed, expanded, gave heavy rain, and dissipated. This time, the cause probably arose from temporary juxtaposition of northern and southern hemisphere vertical circulations. At other times different causes might be adduced.

3.10   Transequatorial Flow

In the monsoon months, over the western Indian Ocean, lower-tropospheric air flows persistently across the equator from the winter hemisphere (high pressure) to the summer hemisphere (low pressure) (Plates I and III). Occasionally in the Northern Hemisphere winter, transequatorial flow prevails over Indonesia and may dominate eastern Africa (Johnson and Mörth, 1960) where it is labeled "cross-equatorial drift" (Fig. 3.33).

As Plates I and III show, the flow has an easterly component in the winter hemisphere and a westerly component in the summer hemisphere, curving most just across the equator in the summer hemisphere. Weather is fair along the equator, but is quite unsettled between the equator and 5° of latitude in the winter hemisphere and around 15° of latitude in the summer hemisphere. The bad weather is worse in the summer hemisphere, where it is said by some to mark the Intertropical Convergence Zone.

Superficially, there seems to be no reason for such large weather gradients or for the worst weather to coincide with anticyclonically curving flow (see Johnson, 1964b).

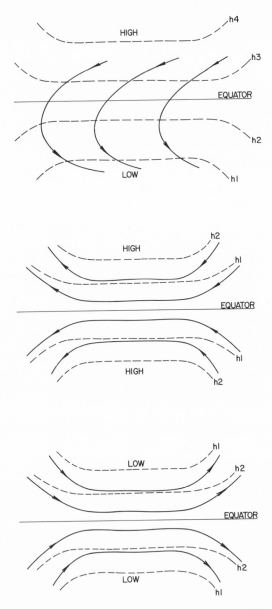

Fig. 3.33. Schematic representations of pressure distribution and flow associated with three equatorial circulation patterns (Johnson and Mörth, 1960). Full lines represent stream lines, dashed lines represent pressure-height contours. (Top) cross-equatorial "drift"; (middle) equatorial "duct"; (bottom) equatorial "bridge."

3.10.1   PARTICLE DYNAMICS

3.10.1.1   *Frictionless Flow.* Johnson and Mörth (1960) applied particle dynamics to explain the bad weather in the summer hemisphere over tropical Africa.

Starting with the equations of motion for horizontal frictionless flow in cartesian coordinates:

$$du/dt = \quad fv - (1/\rho)(\partial p/\partial x) \tag{3.6}$$

$$dv/dt = -fu - (1/\rho)(\partial p/\partial y) \tag{3.7}$$

they assumed a stationary meridional pressure gradient $(\partial p/\partial x = 0)$ and approximated

$$f \approx 2\Omega y/a \tag{3.8}$$

Then

$$du/dt = 2\Omega yv/a \tag{3.9}$$

and through integration

$$u_1 = u_0 + \Omega(y^2 - y_0^2)/a \tag{3.10}$$

and so

$$dv/dt = (-2\Omega y/a)[u_0 + \Omega(y^2 - y_0^2)/a] - (1/\rho)(\partial p/\partial y) \tag{3.11}$$

which may be integrated after multiplying by v

$$v_1 = \mp[v_0^2 - (\Omega^2/a^2)(y^2 - y_0^2)^2 - (2\Omega u_0/a)$$
$$\times (y^2 - y_0^2) - (2/\rho)(\partial p/\partial y)(y - y_0)]^{1/2} \tag{3.12}$$

The trajectories were then computed using finite difference equations in which

$$\Delta t = t_1 - t_0 \tag{3.13}$$

$$x_1 = x_0 + \tfrac{1}{2}(u_0 + u_1)\,\Delta t$$
$$= x_0 + \tfrac{1}{2}[2u_0 + \Omega(y^2 - y_0^2)/a] \tag{3.14}$$

$$y_1 = y_0 + \tfrac{1}{2}(v_0 + v_1)\,\Delta t$$
$$= y_0 + \tfrac{1}{2}\{v_0 \mp [v_0{}^2 - (\Omega^2/a^2)(y^2 - y_0{}^2)^2$$
$$- (2\Omega u_0/a)(y^2 - y_0{}^2) - (2\partial p/\partial y)(y - y_0)]^{1/2}\} \qquad (3.15)$$

In this case the trajectories are functions of y and the initial conditions only. Despite this, Johnson and Mörth found that convergence along the trajectory, which crossed the equator into the Southern Hemisphere, coincided with the usual location of bad weather between 10 and 15S. Gordon and Taylor (1970) pointed out that initial conditions were so critical that the air parcel trajectory delineated by Johnson and Mörth differed widely from other trajectories which would have resulted had the initial velocity assigned to the air parcel been changed by amounts within the range of normal observational error. The reason lies in the fact that any initial error in wind velocity persists throughout the subsequent trajectory computations.

3.10.1.2 *Flow with Friction.* Gordon and Taylor (1970) also used particle dynamics to study air motion in low latitudes, but avoided some of the problems of Johnson and Mörth's formulation. They confined their attention to surface winds over the ocean, and determined a coefficient of surface resistance $F \approx 2.5 \times 10^{-5}$ sec^{-1} acting against the wind where the frictional force $= FV$. To determine coordinates x, y of a trajectory as a function of time, they derived the following equations. Assuming (1) steady-state pressure fields, (2) coriolis parameter constant over each 1-hr time interval, and (3) friction opposing the motion:

$$x_1 = (u_0 - M)(f^2 + F^2)^{-1}e^{-Ft}(f\sin ft - F\cos ft)$$
$$- (v_0 - N)(f^2 + F^2)^{-1}e^{-Ft}(F\sin ft + f\cos ft)$$
$$+ Mt + [F(u_0 - M) + f(v_0 - N)](f^2 + F^2)^{-1} + x_0 \qquad (3.16)$$
$$y_1 = (v_0 - N)(f^2 + F^2)^{-1}e^{-Ft}(-F\cos ft + f\sin ft)$$
$$+ (u_0 - M)(f^2 + F^2)^{-1}e^{-Ft}(F\sin ft + f\cos ft)$$
$$+ Nt + [F(v_0 - N) - f(u_0 - M)](f^2 + F^2)^{-1} + y_0 \qquad (3.17)$$

where

$$[F(\partial p/\partial x) + f(\partial p/\partial y)](f^2 + F^2)^{-1} = M, \qquad (FM - \partial p/\partial x)f^{-1} = N$$

Horizontal divergence, following the motion, is given by

$$\nabla \cdot \mathbf{V}\,dt = dA/A = \log A_2/A_1 = \log A_n/A_{n-1} \qquad (3.18)$$

where A_n/A_{n-1} is the ratio of the contraction or dilation of the area bounded by three or more adjacent parcels during a prescribed interval of time (usually 1 hr).

Whereas an error δV_0 in the initial value of the wind velocity persists unchanged with time in the case of frictionless flow (Eqs. 3.14 and 3.15), Eqs. 3.16 and 3.17 show that the error is reduced exponentially owing to the effect of friction, and at time t, assuming the sin and cos terms $= 1$,

$$\delta V_t \approx e^{-Ft} \delta V_0 \qquad (3.19)$$

Under average conditions in low latitudes the effect of an initial wind velocity error would be reduced to one-third in about 10 hr. Thus, although the equations used by Gordon and Taylor are much more complex than those of Johnson and Mörth (1960) and require calculations to be made on an electronic computer, they are significantly less sensitive to initial-value errors.

3.10.1.3   *Sample Trajectories and Divergence Computations.* Gordon and Taylor simulated release of air parcels along 60E between 15N and 15S and computed their subsequent trajectories, using January long-term mean meridional pressure gradient and mean resultant winds at the release points as input data. Figures 3.34, 3.35, and 3.36 reveal encouraging agreement between computed and climatological winds (indicating that the frictional estimation was reasonable) and more importantly, between divergence and cloudiness.

Figures 3.37 and 3.38 compare the divergence of the computed surface-wind trajectories and observed rainfall frequency over the Arabian Sea for July 1963. Once again, agreement is encouraging, at least in the equatorial zone.

Along the trajectories, acceleration, coriolis, friction, and pressure-force terms combine to determine the magnitude and sign of the divergence. In the January case, Gordon and Taylor found that the distribution of divergence within 5° of the equator was largely determined by the divergence of the pressure-force term (or the curvature of the pressure profile, since $\partial p/\partial x = 0$); farther poleward the coriolis terms predominated. Thus, in the equatorial zone, accurate, *representative* pressure measurements are essential to successful attempts quantitatively to relate the force and weather fields. Random errors prevented Gordon and Taylor from achieving such a goal synoptically (Section 6.1.1.1); they succeeded only when pressures could be averaged over at least several days during which pressure and weather distributions changed insignificantly. Thus day-to-day weather variations in transequatorial flow can seldom be related to variations in the synoptic fields of surface pressure and surface wind.

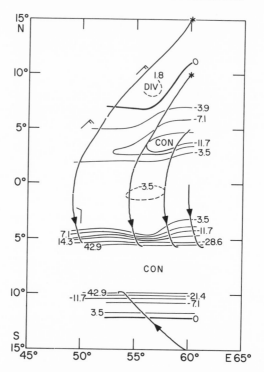

Fig. 3.34. Computed surface trajectories and divergence in units of 10^{-6} sec^{-1} for January in the Indian Ocean. The pressure field was taken as the long-term meridional profile; initial winds equalled the long-term mean resultant winds (Gordon and Taylor, 1970).

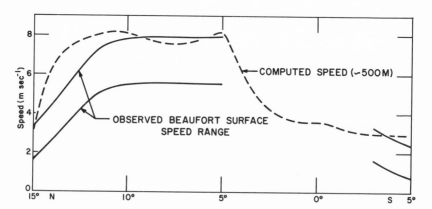

Fig. 3.35. Comparison of observed and computed speeds (m sec^{-1}) along the trajectory released at 15N (Fig. 3.34). The two full lines represent the speed interval corresponding to the Beaufort code number along the trajectory. The region between 5N and 3S had insufficient data (Gordon and Taylor, 1970).

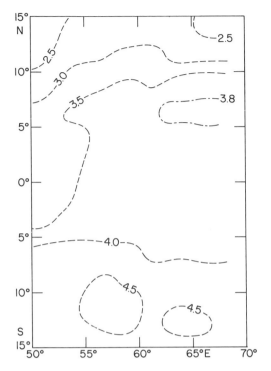

Fig. 3.36. Average cloudiness (oktas) observed from satellites for January 1966 and 1967 (Sadler, 1969).

3.11   Squall Lines

Squall lines are generally defined as nonfrontal lines of active thunder-storms, several to some tens of kilometers wide and hundreds of kilometers long, which exist for a considerably longer period than the lifetime of the component cumulonimbus elements.

Although squall lines are sometimes classified as mesoscale phenomena, studies of them in the monsoon area have largely depended on analyses and descriptions based on synoptic observations. However, with the advent of radar, squall lines have been mapped, tracked, and analyzed and considerable statistics accumulated, especially over the United States, India, and western Africa.

Many individual cumulonimbus in a squall line have lifetimes of an hour or less, although for favored individuals this may be longer. Thus, for a squall line to persist several hours, new convective elements must continually replace dying cells. Although squall lines usually develop or become most intense in

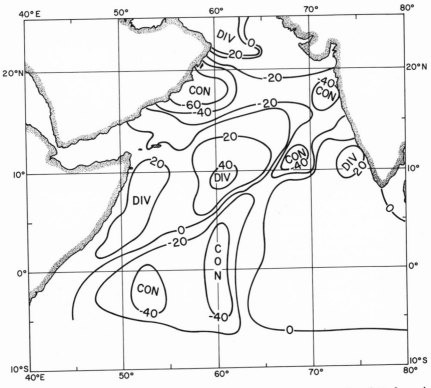

Fig. 3.37. Computed divergence of mean resultant surface winds in units of 10^{-6} sec^{-1} for July 1963 (Gordon and Taylor, 1970).

the late afternoon, they can persist throughout the night when isolated cumulonimbus have long since dissipated. Squall lines, then, behave differently from the sum of their convective parts.

Hamilton and Archbold (1945) described West African squall lines which develop and move westward in an environment of moist lower tropospheric southwesterlies and relatively dry upper tropospheric easterlies.

Rising warm moist air enters the base of the cloud near its leading edge (Fig. 3.39). The updraft extends in the upshear direction. Thus, as the air cools, water vapor condenses and rain falls both from the main cloud column and an extensive overhanging mass. Air originating in the middle troposphere *behind* the storm sinks as it overtakes the storm. Moving beneath the overhanging cloud it is so cooled by evaporation from rain that its leading edge resembles a cold front, displacing and lifting the warm moist air in the path of the storm.

In this section I propose to discuss the mechanism of squall-line generation,

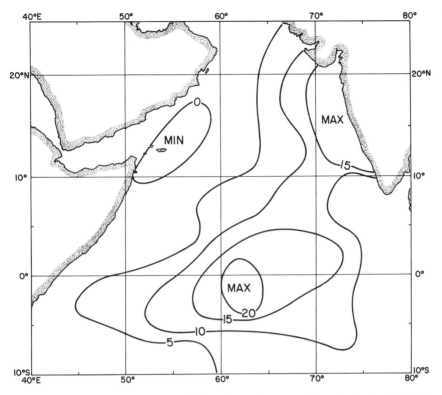

Fig. 3.38. Percentage frequency of ships observations reporting rain during July 1963 (Gordon and Taylor, 1970).

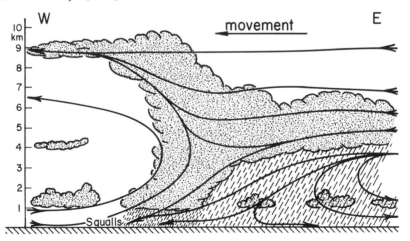

Fig. 3.39. Instantaneous circulation (streamlines) typical of a vigorous West African squall line (based on Hamilton and Archbold, 1945; Tschirhart, 1959).

regeneration, and movement, based on the excellent treatment by Newton
and Newton (1959), and then to describe two forms of their model most
commonly observed in the monsoon region.

A storm system, defined as one squall-line segment comprising several
cumulonimbus, is represented schematically in Fig. 3.40. According to Newton (1967),

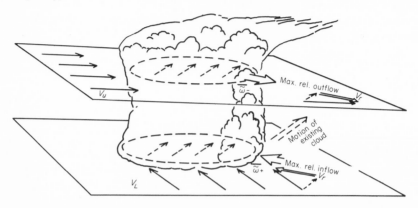

Fig. 3.40. A thunderstorm embedded in a wind field in which the wind veers with height.
Solid arrows denote winds outside storm, at levels near its base and top. Dashed lines denote
approximate average winds inside storm column, $\tilde{\omega}+$ and $\tilde{\omega}-$ denote positive and negative
nonhydrostatic pressure (Newton, 1960).

" Momentum characteristic of the lower levels is carried up in the updrafts.
In the upper part of the cloud, the air tends to take on the momentum of the
upper levels, which in turn is transported earthward in the downdrafts. As a
result of vigorous vertical mixing, the vertical shear within the storm is
partially annihilated, and the in-cloud air at both upper and lower levels
assumes a mean velocity intermediate between the ambient wind velocities
at upper and lower levels.

As a consequence, the motion V_r of the outside air with respect to the
in-cloud air is as indicated by the broad arrows. In relative motion, the
lower-level air (comprising the moist layer) blows into the right flank of the
storm, while the upper-level air blows away from that flank."

Although air is continually exchanged between storm and environment,
the storm impedes the environmental flow. This induces a nonhydrostatic
pressure

$$\tilde{\omega} = \tfrac{1}{2}(\rho V_r{}^2) \tag{3.20}$$

which is positive where V_r is directed toward the storm, and negative where
V_r is directed away from the storm.

3.11.1 MECHANISM OF THE SQUALL LINE

The Newtons determined that

$$\mathbf{V}_r^2 \propto D(w\,\partial C/\partial z) \tag{3.21}$$

whence it can be seen that the broader the storm and the more vigorous the vertical motion, the less is the storm distorted by environmental shear and the greater is $\mathbf{V}_r$ and, consequently, $\tilde{\omega}$.

The Newtons then went on to point out how the nonhydrostatic pressure field induced by the relative motions of air within and surrounding the storm aids convection, and derived an approximate relation

$$dw/dt \approx g(\Delta T/T_e + \partial\tilde{\omega}/\partial p) \tag{3.22}$$

where the first term on the right is the buoyancy force and the second the vertical gradient of the nonhydrostatic pressure. The term $\partial\tilde{\omega}/\partial p$ depends on the difference between $\tilde{\omega}$ at the base and $\tilde{\omega}$ at the top of the storm. Since nonhydrostatic pressures induced on the storm "cylinder" are proportional to $\rho\mathbf{V}_r^2$, Eq. (3.21) shows that the greater the vertical wind shear in the environment and the greater the size of the storm, the greater is $\partial\tilde{\omega}/\partial p$. Because vertical shear within the storm may be less than half of the environmental shear, maximum $\partial\tilde{\omega}/\partial p$ occurs on the down-shear side of the storm, with nonhydrostatic pressure positive at the base and negative at the top. Here convective elements continually build, whereas on the up-shear side $\partial\tilde{\omega}/\partial p$ is negative and downward motion dissipates the clouds. Thus the storm "moves" in the direction of the developing elements and away from the dissipating elements. Over the United States and during spring over India, winds veer with height from south to west. With such a distribution, Fig. 3.40 shows that maximum $\partial\tilde{\omega}/\partial p$ is found to the right of the mean wind within the cloud. The storm movement reflects a combination of the movement of individual convective cells and the regenerative process resulting from the nonhydrostatic pressure effect.

Newton and Katz (1958) studied storms of this type over the central United States. On the average, the storms moved 25° to the right of and 4 m sec^{-1} slower than the mean 700-mb wind. Boucher and Wexler (1961) found that squall lines over the midwestern and northeastern United States, respectively, exhibited modal directions of movement of about 50 and 70° to the right of the 700-mb wind. In both regions low-level flow from a southerly quarter ensured a plentiful supply of moisture.

These results are not necessarily divergent since squall *lines*, comprising numerous storms, tend to move to the right of the directions followed by individual storms.

The cold low-level outdraft from the storm (see Fig. 3.39) tends to enhance both the nonhydrostatic effect (by increasing V_r) and the buoyancy effect (by lifting stagnant air on the storm periphery). Obviously the buoyancy effect [first term on right, Eq. (3.22)] is greatest when the storm ingests air warmed by the afternoon sun. Even at night, when this term may become negative for air in the lower levels, the nonhydrostatic pressure force, by lifting surface air to the free convection level, might maintain the storm, though at reduced intensity.

A squall line, being a chain of storms, behaves as a combination of these components, in contrast to the way individual cumulonimbus behave. Thus, strong vertical shear increases the vertical gradient of nonhydrostatic pressure and strengthens the squall line, but, by enhancing entrainment, inhibits development of an isolated cumulonimbus.

Squall lines usually develop in the afternoon when thermal convection is greatest, in the vicinity of some topographic discontinuity, such as a range of mountains or a valley, which presumably counteracts the initially inhibiting effect of vertical shear. Squall lines usually dissipate at night or when nearing or crossing the coast. In both cases cool, relatively stable surface-air layers prevent buoyant ascent.

In Eq. (3.22), the buoyancy term controls development and contributes to dissipation, whereas the nonhydrostatic pressure term, becoming effective only after a cumulonimbus "factory" has grown, acts powerfully to extend, intensify, and maintain the resulting squall line. In a numerical study, Takeda (1966) confirmed that although strong vertical shear does not of itself intensify a convective system, it effectively prolongs the life of a sufficiently large and vigorous system.

3.11.2   SQUALL LINES IN THE MONSOON AREA

Squall lines similar to those of the United States occur during spring and autumn over northern India (Suckstorff, 1939; Desai, 1950; Newton, 1950; Das et al., 1957; De, 1963), and chiefly in spring over northern Australia (Brunt and Mackerras, 1961; Clarke, 1962). The lines develop when a lower-level moist poleward-flowing current is overlain by vigorous westerlies. Instability is readily realized, vertical shear intensifies and maintains the line, which usually travels eastward and equatorward.

Over central and western Africa, squall lines resemble those over the United States in the following ways: They tend to develop during the afternoon (Hamilton and Archbold, 1945); they are usually most intense during the afternoon (Jeandidier and Rainteau, 1957); and they are most frequent in the spring (Hamilton and Archbold, 1945) when they are favored by a combination of instability and lower-tropospheric moisture (Fig. 5.16).

On the other hand, central and West African squall lines differ in the following ways: They are usually embedded in an environment possessing easterly vertical shear and move westward (Hamilton and Archbold, 1945) at speeds up to 10 m sec^{-1} greater than the environmental 3-km winds (Eldridge, 1957), and move faster in the afternoon than at night (Tschirhart, 1959). Although about 80% last less than 24 hr, a few are relatively long lived, sometimes lasting several days (Cochemé and Franquin, 1967) and traveling more than 3000 km (Eldridge, 1957). Intense squall lines move faster than weak; slowing accompanies degeneration (Tschirhart, 1959). Not all squall lines weaken on crossing the coast, particularly over the warm Guinea current (Hamilton and Archbold, 1945; Eldridge, 1957). These differences are significant enough to demand modification of the Newtons' model; in what

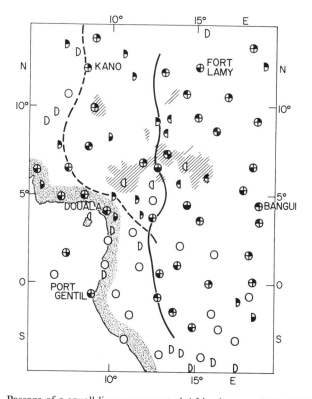

Fig. 3.41. Passage of a squall line across central Africa between 0600 GMT (solid line) and 1500 GMT (dashed line), 3 September 1957. Station circle code: left semicircle—0600 GMT, right semicircle—1500 GMT; blocked-in upper semicircle—thunderstorm, blocked-in lower semicircle—other precipitation. Lagos (6°35′N, 3°20′E) and Abidjan (5°15′N, 3°56′W) as well as the five rawin stations labeled on the chart were used in the calculations. Land above 1000 m hatched (from Tschirhart, 1958).

follows, I have drawn heavily on data contained in two papers by Tschirhart (1958, 1959).

3.11.2.1   *Example*. Between 0600 and 1500 hr GMT 3 September 1958, a squall line oriented north–south moved westward across central Africa (Fig. 3.41). Those unfamiliar with tropical African weather would have difficulty delineating squall lines on a synoptic weather chart. In this example, discussed already by Tschirhart (1958), I have depended on his analysis to locate the squall line.

The following computations are based on 0600 GMT rawin observations from seven stations in the vicinity of the squall line. The environmental wind V_e at each pressure level, was the vector mean of the seven station winds at that level (Fig. 3.42). The mean wind acting on the squall line may (Fankhauser, 1964) be approximated by

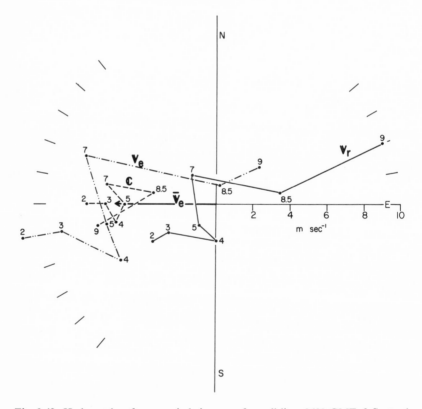

Fig. 3.42. Hodographs of upper winds in area of squall line, 0600 GMT, 3 September 1957, computed from the seven stations of Fig. 3.41. environmental wind, V_e = mean cloud movement; in-cloud wind, C; relative wind, $V_r = V_e - C$. Pressure levels are labeled in decibars.

$$\mathbf{V}_e = (\mathbf{V}_{e850} + \mathbf{V}_{e700} + \mathbf{V}_{e500} + \mathbf{V}_{e300})/4$$
$$= 090° \quad 5.5 \quad \text{m sec}^{-1} \tag{3.23}$$

Assuming an in-storm vertical shear of $\frac{1}{3}$ of the environmental vertical shear leads readily to an estimate for each pressure level of in-cloud wind ($\mathbf{C}$) and of the wind relative to the cloud, $\mathbf{V}_r = \mathbf{V}_e - \mathbf{C}$. Then, following the Newtons and assuming that the elements of the squall line are cylindrically shaped storm systems, which act to some degree as inpenetrable barriers, I computed the nonhydrostatic pressure $\tilde{\omega}$ around the cylinders (Goldstein, 1938). In Fig. 3.43 a storm "cylinder" is collapsed onto a plane with pressure height as the r coordinate. Upward acceleration ($\partial\tilde{\omega}/\partial p > 0$) is largely confined to the WSW segment while ENE of the storm center, downward acceleration ($\partial\tilde{\omega}/\partial p < 0$) prevails throughout the troposphere. $\partial\tilde{\omega}/\partial p$ is large at low levels where it can be effective in helping lift air in the planetary boundary layer to

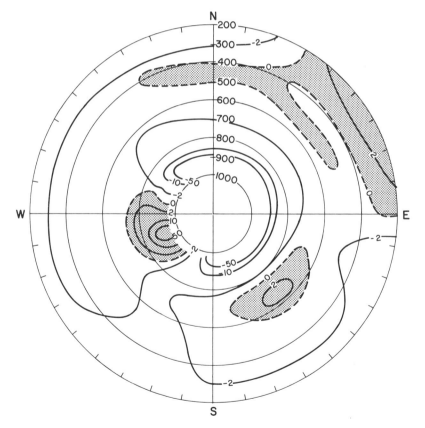

Fig. 3.43. Nonhydrostatic pressure at 0600 GMT, 3 September 1957 in hundredths of mb around the squall line of Fig. 3.41. The r coordinate gives pressure height in millibars.

the condensation level. Beyond this level, ordinary buoyancy forces come into play.

If we assume that the individual convective cells moved at the velocity of the environment $(\mathbf{V_e})$, then, propagation due to the nonhydrostatic term acting along the same line accounts for the storms comprising the squall line moving even faster westward than the 5.5 m sec^{-1} computed above (15 m sec^{-1} at 10N; 12 m sec^{-1} at 5N; stationary and dissipating south of 5N where upper-tropospheric westerlies predominated). The faster such a squall line advances, the more vigorously will air be uplifted by the cold surface outdraft along its leading edge and the more effectively will the buoyancy term $\Delta T/T_e$ enhance the intensity.

These conclusions are in agreement with Tschirhart (1959) who stated that forward movement of a squall line is the sum of two components, speed of development and speed of the environmental flow.

Little net turning of wind with height is typical of central and West Africa. Easterly shear is concentrated in lower and upper tropospheres, with slight or variable shear in the middle troposphere. An increase in easterly shear accompanies development and subsequent movement of squall lines, but, according to Nevière (1959), it is sometimes hard to determine which is cause and which is effect.

A northwestward-moving squall line has been observed near Calcutta during the summer monsoon (De et al., 1957); near the end of 1966 summer monsoon, Bombay radar tracked squall lines which developed over the Western Ghats and then moved westward to more than 200 km beyond the coast (Argawal and Krishnamurthy, 1969). However, squall lines occur only rarely in summer monsoon *rains* regimes because an environmental lapse rate close to saturated adiabatic, combined with vigorous easterly vertical shear inhibit initial thunderstorm development (Section 4.1.1), and also prevent generation of a strong cold downdraft.

... on this Coast of Surat and Malabar, when the Rains are over, keep exactly Land-Breezes from Midnight to Mid-day, and Sea-Breezes from the Noon of Day to the Noon of Night.

John Fryer, *A New Account of East-India and Persia*, 1698.

4. Precipitation and Mesoscale Features

4.1 The Character of Precipitation in the Monsoons

In Section 2.5.2 I introduced the terms "*rains*" and "*showers*." In this section I discuss geographical distribution and annual variations of *rains* and *showers*, their differing dynamic and thermodynamic properties, and their associations with synoptic-scale circulations.

4.1.1 THUNDERSTORM FREQUENCY AND RAINFALL

Heavy rain falls from thunderstorms. The length of the average thunderstorm life is about the same everywhere. Consequently, where thunderstorms are the overwhelmingly predominant rain producers, areal variation in the ratio of the average number of thunderstorm days to the average rainfall should be slight. Figure 4.1 derived from Portig (1963), shows large areal variation. The areal ratio ranges from 16 over western Africa and southeastern Tibet to less than 2 over China, southwestern India, and the ocean. At the ends of the range Portig identified two types of monsoon rain regime—the *West African*, in which frequency of thunderstorm days and rainfall increase to maxima in midsummer; and the *western Indian* in which frequency of thunderstorm days decreases as the rainfall increases.

Indian meteorologists have long known that periods of maximum thunderstorm frequency precede and follow the summer *rains* (Ram, 1929). The *rains*, particularly along the west coast of India, are notable for a dearth of thunderstorms.

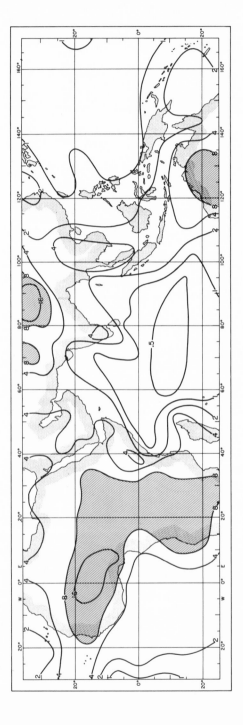

Fig. 4.1. Ratio of the annual mean number of thunderstorm days to the mean annual rainfall in decimeters (from Portig, 1963).

Between Bombay and northern India, where four rawinsonde stations are located, the thunderstorm days/rainfall ratio spans both of Portig's categories (Fig. 4.1). In an attempt to analyze the gradient, I compared monthly averages of the ratio, rainfall, and the lapse rate and vertical wind shear between 850 and 300 mb at the stations (Fig. 4.2).

Jodhpur and New Delhi, with most thunderstorm days in July are in Portig's West African category; at Bombay and Ahmedabad, archetypal of the western Indian category, thunderstorm days are fewest in August. However, at *all four stations* the thunderstorm days/rainfall ratios are least in midsummer when rainfall is greatest. From May to August at Bombay, where the *rains* are heavy, the ratio decreases from 6.5 to 0.06, whereas at Jodhpur on the edge of the Great Indian Desert, the decrease is an order of magnitude less. During the *rains* the ratio is an order of magnitude less in the south than in the north.

Thus in both categories the trend toward midsummer is the same. Average thunderstorm raininess is unlikely to increase spectacularly. Perhaps then, a different mechanism begins to produce rain, becoming quite important in the north and overwhelmingly predominant in the south.

In the *south*, as summer advances, the monsoon circulation develops and intensifies. Strong and generally convergent westerlies underlying strong and generally divergent easterlies, by leading to large-scale upward motion through the middle troposphere, raise humidities, decrease lapse rates to near-moist adiabatic, and develop great cloud sheets with embedded cumulonimbus, from which considerable rain falls. The reduced insolational heating, near-stable lapse rate, and large vertical wind shear all inhibit thunderstorm formation, and a *rains* regime prevails.

In the *north*, the intensifying monsoon by deepening the moist layer decreases the lapse rate and hinders thunderstorm development. However, Jodhpur and New Delhi being near the heat trough and upper-tropospheric ridge axes, experience decreasing vertical wind shear, a trend which favors thunderstorm development.

During summer the upper troposphere is relatively dry, the lapse rate is conditionally unstable,* shear is small, and insolational heating is favored —ideal for thunderstorm development, but not for extensive rain. A prevailing *showers* regime only rarely gives way to *rains*. Meridional profiles over western Africa are similar (Fig. 5.16).

The impression that frequency of thunderstorm days and rainfall are inversely related is reinforced by mesoscale studies in Southeast Asia.

---

* Because 850-mb temperatures are much higher in the north than in the south, the difference in conditional instability between north and south is greater than suggested by the curves.

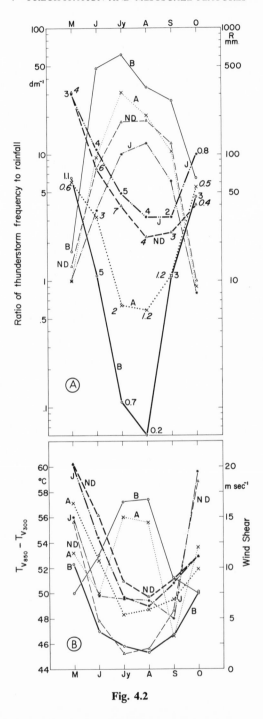

Fig. 4.2

Summer vertical profiles of virtual equivalent potential temperature at Southeast Asian stations reveal greater convective instability on fine days than on wet days (Fig. 4.3) (Harris *et al.*, 1970). Valovcin (1970) reported an inverse relationship between the wind speed at 850 mb and the number of radar precipitation echoes above 13.7 km within 95 km of Saigon. Near

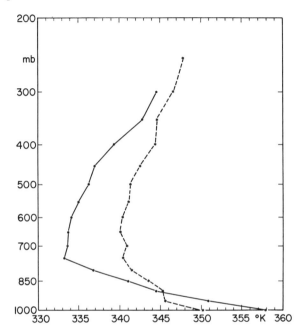

Fig. 4.3. Vertical profiles of mean virtual equivalent potential temperature (θ_{ve}) at 1200 GMT for Saigon (10°49′N, 106°40′E) during periods of good weather (full line) and bad weather (dashed line) during July 1966 (Harris *et al.*, 1970).

Fig. 4.2. Monthly means of thunderstorms, rainfall, lapse rate, and vertical wind shear at four stations in western India:

Bombay	19°07′N, 72°51′E	Full lines
Ahmedabad	23°04′N, 72°38′E	Dotted lines
Jodhpur	26°18′N, 73°01′E	Dot-dashed lines
New Delhi	28°35′N, 77°12′E	Dashed lines

(a) Ratio of thunderstorm frequency to rainfall in decimeters (heavy lines), where numbers denote thunderstorm frequencies; rainfall (light lines). (b) Difference in virtual temperature between 850 and 300 mb (heavy lines), at the moist adiabatic lapse rate the difference ranges between 33.5C ($T_{850} = 29C$) and 43.5C ($T_{850} = 20C$); wind shear between 850 and 300 mb (light lines).

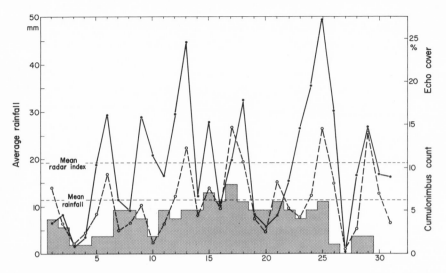

Fig. 4.4. Rain and clouds in the Saigon area during July 1967 (Harris *et al.*, 1969; Valovcin, 1970): percentage of the area covered by radar precipitation echoes above 6 km and between noon and midnight (full line); average daily rainfall at 14 stations (dashed lines); number of cumulonimbus radar echoes above 13.7 km (histogram).

Saigon (Fig. 4.4) the daily count of cumulonimbus above 13.7 km is poorly correlated with the percentage area covered by radar echoes above 6 km and correlated not at all with 24-hr rainfall totals.

Thus a paradox emerges. During the summer monsoon, extensive fine weather is also thunderstorm weather. Beneath scattered convective cells, showers may be heavy but average rainfall per unit area is only a fraction of that caused by the *rains.*

Gentle subsidence favors fine weather, since the accompanying inversion or isothermal layer limits convection. However, an inversion can also *intensify* convection (Fulks, 1951). Air beneath the inversion is moist and becomes warmer in response to daytime heating; air above the inversion may become cooler. And so potential instability throughout the troposphere increases until some thermal or orographic agent ruptures the inversion. Then thunderstorms develop.

4.1.2   Rains and Showers in Synoptic-Scale Circulation

Charney (1967), suggested that formation of a warm-core tropical depression (in which *rains*-type precipitation predominates) results from

" a cooperative, frictionally controlled interaction between an ensemble of cumulus cells and a large-scale cyclonic disturbance [see Fig. 4.5]. In

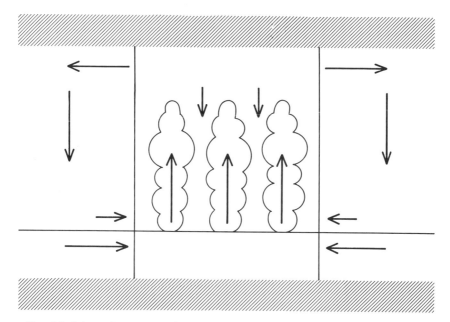

Fig. 4.5. Schematic representation of interaction between cumulus cells and a large-scale cyclonic disturbance (Charney, 1967). The picture would probably be improved by isolated towering cumulus in the descent regions.

the region of cyclonic vorticity of the larger-scale field of motion, frictionally induced convergence at low levels causes the formation of a deep moist layer which supplies the energy for deep cumulonimbus development."

Synoptic scale convergence in the friction layer enhances total rainfall through massive ascent but, by increasing the depth of moist air and diminishing the lapse rate, hinders thunderstorm development. Conversely, sinking motion outside the storm circulation reduces the quantity of precipitable water but, by increasing the lapse rate, creates an environment favoring scattered thunderstorm development.

Friction-layer convergence leads to *rains* in the circulations of tropical storms and monsoon depressions and in the accelerating westerlies south of these systems (Sections 3.3 and 3.4). In winter, when the monsoon freshens, the windward coasts of Indochina and the Philippines disappear beneath palls of nimbostratus. Thundershowers are rare.

In subtropical cyclones (Section 3.5), with maximum vorticity and convergence in mid-troposphere, the role of friction-layer convergence in spreading moisture through the troposphere is much less than in the case of warm-core depressions. However, moisture is spread efficiently by evaporationally cooled downdrafts. Vertical gradients of θ_e are small (Fig. 3.10), *rains* predominate, and the rare thundershowers usually develop outside the main circulation (*frontispiece*).

Rains fall in association with accelerating surface westerly winds equatorward of typhoons, monsoon depressions, or subtropical cyclones, or near the summer polar front of Southeast Asia (Sections 5.8.4.1 and 5.8.6.2). Overland thunder*showers* occur before and after the summer monsoon, near the heat trough (Fig. 5.42) during breaks in the *rains*, and in the shadow of mountain ranges during the *rains* (see below). Over the sea, on the other hand, thunderstorms are rare in these situations (La Seur, 1967) because insolational heating is so much weaker than over land.

In subsequent discussion of synoptic sequences I call attention to the two precipitation types, which are readily identifiable in weather satellite pictures.

4.1.3 DIURNAL VARIATION OF RAINFALL

Throughout the monsoon area orography and shoreline orientation act powerfully to change the patterns of vertical motion, and hence of the *rains* and *showers* mentioned in the preceding paragraphs. Local winds, caused by mesoscale diurnal variations in heating gradients, may produce bewilderingly complex diurnal variation of rainfall patterns. Any serious attempt to explain these patterns demands that, first, diurnal variations in the three-dimensional fields of heat, motion, and moisture be studied.

Three-hourly serial rawin soundings made for several weeks over a mesoscale network might provide sufficient data for a single study. But such information is unavailable anywhere, let alone in the monsoon area, and so I confine myself to giving some scattered hazy glimpses of a problem in search of observations.

Nicholson (1970) pointed out that the circulation over the Indian subcontinent varies diurnally and out of phase with the sea-breeze–land-breeze regime cycle. Interaction between the two is therefore complex, resulting, along the west coast, in some hours delay in sea-breeze and land-breeze onsets, a phenomenon which Fryer accurately observed almost 300 years ago. Kreitzberg (1970), studying averages of 3-hourly radiosonde observations in the Saigon area, found that middle tropospheric specific humidity undergoes a diurnal cycle. A minimum at noon and a maximum at midnight he explained as being due to subsidence and ascent, respectively, resulting from diurnal variations in the pressure gradient over the interior of the peninsula. Three-hourly radiosoundings have not been made elsewhere in the monsoon area and so it is hard to check whether Kreitzberg's finding is widely valid during the summer monsoon. Averages of 6-hourly soundings made at Taipei (25°04′N, 121°32′E) (Ling, 1965) and of 12-hourly soundings at

Hong Kong (22°19'N, 114°10'E) (Royal Observatory, 1961) show no significant diurnal variation in specific humidity, while I could detect no trend between 7- and 13-hr soundings at Khartoum (15°36'N, 32°33'E), near the axis of the heat trough.

We must depend therefore on little more than intuitive physical explanations of diurnal rainfall variations.

4.1.3.1 *Showers Regime.* Predominantly clear skies enhance mountain/valley and sea-breeze/land-breeze circulations which, in turn, by acting on a relatively unstable air mass, favor development of showers and thundershowers along the foothills during the afternoon and offshore at night (Weldon, 1969; Bell, 1970).

In the equatorial doldrums these local circulations overcome weak monsoon or fitful winds and control vertical motion (Braak, 1940; Hamilton and Archbold, 1945; Thompson, 1957b; Ramage, 1964a; Nieuwolt *et al.*, 1967) (Fig. 4.6). They are modulated but not cancelled by synoptic changes.

For example, in ten years, rain was *never* recorded during August at Kota

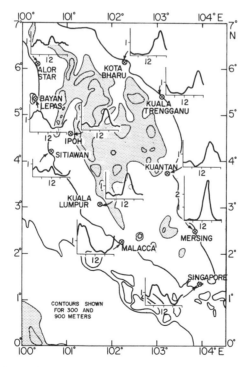

Fig. 4.6. Diurnal variation of August rainfall over Malaya: 3-hourly running means in centimeters for 11 stations; period of record 1950–1959 (Ramage, 1964a).

Bharu (6°10'N, 102°17'E) between 0900 and 1100 hr although the monthly average amounted to 146 mm. At Kisumu (0°06'S, 34°45'E), on the eastern shore of Lake Victoria, none of the average March rainfall of 160 mm falls between 0800 and 1400 hr. Not all tropical stations are forecasters' nightmares! Rain shadowing is unimportant. The southwest monsoon blows steadily *though weakly* across Malaya in August and rainfall in the east, where sea breezes replenish moisture, is as heavy as rainfall in the west.

4.1.3.2 *Rains Regime.* Predominantly cloudy skies and fresh or strong surface winds, by reducing surface heating gradients, diminish mountain/ valley and sea-breeze/land-breeze circulations. A near-moist adiabatic lapse rate handicaps intense heat convection. Except where mentioned below, nocturnal falls are significant and have been reported over Africa (Hamilton and Archbold, 1945; Jeandidier and Rainteau, 1957), southern Asia (Iyer and Dass, 1946; Iyer and Zafar, 1946; Narasimham and Zafar, 1947; Ramaswamy and Suryanarayana, 1950; Rao and Raman, 1959), eastern Asia (Ramage, 1952a), Indonesia (Braak, 1921–1929; Preedy, 1966) and northern Australia (Southern, 1969). It is tempting to try and relate this type of diurnal variation in which considerable rain falls at night to a corresponding diurnal variation in the summer monsoon circulation (Ramage, 1952a), but perhaps the explanation offered by Jeandidier and Rainteau (1957) and earlier suggested by the work of Refsdal (1930) will suffice until detailed studies are made. In a deep, strong convergent monsoon current, considerable cloud persisting through the night and morning inhibits insolation and development of a notable afternoon maximum. At night, radiational cooling from the tops and reduced surface cooling beneath an extensive cloud system presumably increase instability, cloud depth, and rainfall. During the day, radiational warming from the tops and reduced insolation at the surface presumably increase stability and consequently decrease rainfall. However, complications resulting from overturning cannot be evaluated.

Fresh or strong lower-tropospheric winds associated with *rains* act on mountain ranges to distort prevailing diurnal patterns. On the windward slopes uplift increases total rainfall as well as its nocturnal component. Air moving to leeward has been dried by precipitation and descent. Nights are often clear but afternoon thunderstorms are not uncommon, arising from insolation and unstable middle-tropospheric lapse rates—a diurnal sequence typical of a *showers* regime, and observed in the lee of Indonesian ranges (Braak, 1921–1929), over northern Taiwan (Ramage, 1952a), in the environs of Saigon (Weldon, 1969), and probably elsewhere as well.

The monsoon area meteorologist is this obliged first to anticipate synoptic change and then to apply his knowledge of orographic and land–sea effects to forecast local weather.

4.2   Mesoscale Features

The discussions of bad weather have implied that upward motion, condensation, cloud depth and, consequently, rain are rather evenly distributed through systems of synoptic scale. However, radar and weather satellites have revealed considerable spatial variability on a scale larger than a cumulus cloud but smaller than synoptic. Though this *mesoscale* is a feature of all synoptic disturbances, nowhere in the monsoon area are sounding data sufficient to describe it three dimensionally.

4.2.1   CLIMATOLOGY

Rainbird (1968), using a technique developed by Henry and Griffiths (1963), studied the distribution of daily rainfall over the Se San subbasin of the Mekong River, in three classes of summer disturbance: low-level cyclones, subtropical cyclones, and trough lines—all *rains* regimes. Figure 4.7 shows

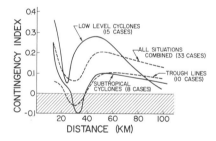

Fig. 4.7. Contingency index as a function of distance between rainfall stations for various types of synoptic disturbances (Rainbird, 1968). Measurements were made over the Se San subbasin of the Mekong River during the summers of 1961 and 1962. The contingency index (Henry and Griffiths, 1963) tests simultaneous occurrence at two stations of 24-hr total rainfall within one of three class intervals: no rain, >25 mm, <25 mm. $CI = 1.0$ when rainfall at a pair of stations is in the same class interval; $CI = 0$ represents random chance.

that the three classes, for which curves of the "contingency index" are plotted against distance, are similar. Maximum values occur at short distances but surprisingly, all curves dip sharply to minima at between 25 and 30 km. Secondary maxima are centered near 50 km.

With low-level cyclones, which generally move slowly westward, the contingency index minimum does not drop below zero. For both subtropical cyclones and trough lines, however, the index minima are negative, indicating that in these heavy rain-producing disturbances, a chance of the same category of rain occurring at stations 25–30 km apart is less than random chance. Rainbird ascribed this to mesoscale processes within the larger synoptic systems.

Since subtropical cyclones and trough lines are usually quasistationary the curves strongly suggest that over the Mekong, mesoscale centers of action are likely to be about 50 km apart and to be themselves quasistationary.

Billa and Raj (1966) used a triangle of recording rain gauges near Poona (18°35′N, 73°55′E) to measure the dimensions of five mesoscale rain systems during the summer of 1964. The median radius was 55 km.

If mesoscale systems so distort *rains* regimes, they must play at least as significant a role in determining local weather in *showers* regimes.

4.2.2   TORRENTIAL LOCAL RAINS

A thunderstorm or squall line may drench a spot for up to 1 hr, although rainfall seldom exceeds 50 mm. Much more damaging are the rare storms which last several hours and from which more than 250 mm falls over a few hundred square kilometers. Their very intensity has handicapped investigation. Studies in India (Iyer and Zafar, 1940), Indonesia (Braak, 1921–1929), and Hong Kong (Bell and Chin, 1968; Chen, 1969), suggested that the storms:

(1) are associated with synoptic-scale disturbances—tropical cyclones, monsoon depressions, strong monsoon winds, or troughs;

(2) draw on a plentiful supply of moisture, either along a coast or from flooded river plains such as the Ganges;

(3) are usually anchored by a discontinuity in surface roughness—generally a coastline or a mountain range.

Torrential local rains stem from two types of storms: those with weak and those with strong surface winds.

4.2.2.1   *Storms with Strong Surface Winds* are caused by tropical cyclones or vigorous monsoon circulations. Convergent lower-tropospheric and divergent upper-tropospheric flow induce massive upward motion. Orographic uplift and slow synoptic change intensify the downpour. As in the *rains* situation (Section 4.1.1), considerable vertical shear inhibits thunderstorm development.

These storms have occurred around the shorelines of the South China Sea and the Bay of Bengal when tropical cyclones began moving inland, and frequently on the seaward slopes of the Cardamon Mountains in Cambodia, the Arakan Yoma in Burma, the Western Ghats and southern Himalayas in India, and the Fouta Djalon in Guinea during periods of vigorous summer monsoon flow. They occur on windward mountain slopes in Indonesia when monsoons are active. At Cherrapunji (25°15'N, 91°44'E), on a hill-ringed plateau, 1300 m up the south face of the Khasi Hills in Assam, orography and the summer monsoon mesh perfectly. Three times every 2 yr, on the average, 24-hr falls exceed 500 mm. Over all the rest of India this amount can be expected only once a year (Iyer and Zafar, 1940).

4.2.2.2 *Storms with Weak Surface Winds* can develop along the polar front or a near-equatorial trough or within the circulation of a dissipating monsoon depression. They are huge, stationary, persistent thunderstorms, apparently initiated and sustained by upper-tropospheric divergence (Sourbeer and Gentry, 1961; Daniel and Subramaniam, 1966; Chen, 1969) above geographically anchored surface convergence. Besides, radiation and advection in the upper troposphere must so rapidly dissipate the heat released by condensation in the rising air that thunderstorm lapse rates are maintained for hours. When these exceptional circumstances supervene, the effects are both spectacular and eerie—lightning flashing within the pallium, an incessant freight-train roar of thunder, and rain sheeting through the breathless air.

Examples. During 18 hr on 12 June 1966 an intense storm on the polar front flooded Hong Kong with up to 600 mm of rain in places and more than 250 mm over an area of 750 km^2 (Chen, 1969) (Figs. 4.8 and 4.9). This rate required an average upcurrent of about 0.5 m sec^{-1}. Thunder and lightning were almost continuous and the tropospheric lapse rate exceeded moist adiabatic by 0.6C 100 mb^{-1}.

At Aberdeen, on the south side of Hong Kong Island, 340 mm fell in a 4-hr, squall-free downpour. Usually entrainment and peripheral sinking would stop the rain much sooner. Here, a great mass of storm clouds protected a considerable inner core against entrainment, while efficient upper-tropospheric heat dispersion sufficiently far from the storm prevented local compensatory subsidence and buoyancy loss.

Hong Kong was enveloped in moist convergent southwesterlies south of a lower-tropospheric depression (Fig. 4.10, bottom). Directly above, winds flowed out from a 200-mb anticyclone (Fig. 4.10, top) both to north (downstream to a jet maximum) and to south, where easterlies exceeded 20 m sec^{-1}. Two powerful vertical circulations were thus linked in a single upward branch over southern China. The northern one was a typical spring circulation in which potential energy of condensation was transformed into a surplus of kinetic energy (Section 5.8.6.2). The southern one typified the summer monsoon (Section 5.8.7.1), with a heat sink over the southern part of the South China Sea.

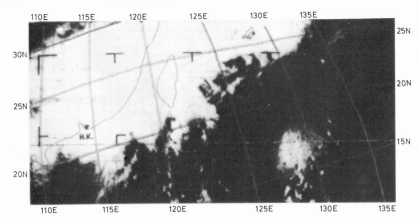

Fig. 4.8. Clouds photographed from ESSA 2 near 0030 GMT, 12 June 1966 (Orbit No. 1312).

Fig. 4.9. Automobiles swept by flood waters down Ming Yuen Street, Hong Kong, on 12 June 1966. Photo courtesy of Hong Kong Government, Department of Information.

Between 1884 and 1966, 24-hr nontyphoon rainfall at the Royal Observatory, Hong Kong equaled or exceeded 250 mm 11 times (Chen, 1969) with the following monthly distribution:

May	June	July	Aug.	Sept.	Oct.
2	7	1	0	0	1

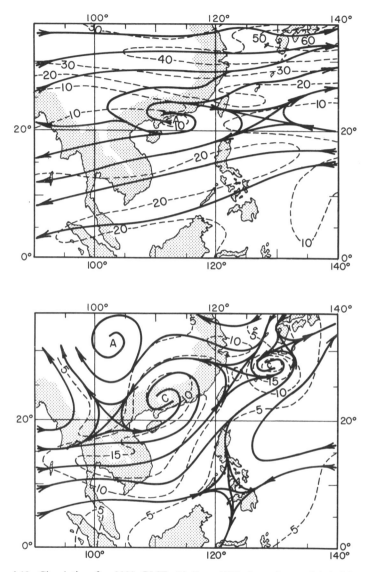

Fig. 4.10. Circulation for 0000 GMT, 12 June 1966. Isotachs are labeled in meters per second. (Top) 200-mb level; (bottom) 850-mb level (Chen, 1969).

Perhaps the immense heat dissipation required to sustain a " continuous " thunderstorm over southern China can be achieved only by means of the interacting vertical circulations shown in Fig. 4.10. If so, the relatively frequent torrential rains in June would not be surprising, because in June southern China lies in the transition zone between the Mei-Yü to the north (Section 5.8.6.2) and the summer monsoon to the south.

South Kerala experienced torrential thunderstorm rain on 17 and 18 October 1964, with Trivandrum Observatory (8°29′N, 76°57′E) recording 401 mm in 10 hr (Daniel and Subramaniam, 1966). The region lay southwest of a surface depression and beneath a 200-mb ridge. From the ridge northward, winds diverged downstream into the newly established winter jet over northern India (Section 5.8.1). From the ridge southward diverging northeasterlies crossed the equator. Where the upward branches of vertical circulations typical of summer and winter linked, the skies opened.

The torrential thunderstorm rain at Kuantan in Malaya (316 mm on 10 March 1967) was associated with a similar transition situation (Section 3.9.3.2).

Giant continuous thunderstorms do not fit the thunderstorm model (Byers and Braham, 1949). Neither can they be categorized as *rains* or *showers*. Determining how energy cascades through them between convective and synoptic scales might more readily achieve a prime objective of the Global Atmospheric Research Program (Bolin, 1967), than measurements of lesser phenomena.

... the processes which give rise to the monsoon in India are very complex and highly inter-related; any disturbance in one affects all the others, and it is impossible to state which is cause and which is effect.

Sir George Simpson, *The South-west Monsoon,* 1921.

5. March of the Seasons

5.1 Role of the Himalayan–Tibetan Massif in the Monsoons

In Chapter 2 I mentioned briefly how the chain of mountains extending from Turkey to western China protects southern Asia from cold outbreaks and so results in a sharp discontinuity in surface monsoon characteristics along about 100E.

Figure 5.1 reveals a large cloudiness gradient between northwestern India and western China, a distribution which prevails throughout the year and which is also reflected in the rainfall.

Recent investigations, especially by Flohn (1968), leave little doubt that this weather discontinuity derives from the mountain–plateau mass of the Himalayas and Tibet which pervasively affects the entire Northern Hemisphere monsoon area throughout the year, and at times even influences climate in the Southern Hemisphere monsoon area (Section 7.1.2).

Thus the role of this great thermal, dynamic pivot should be considered prior to following the march of the seasons.

In the broadest sense, the Northern Hemisphere monsoon area comprises three parts: east of Tibet, where significant winter precipitation falls and spring and summer are wet; west of Tibet, where deserts dominate; and south of Tibet, where winters are desertlike and summers are wet. The 200-mb mean resultant winds reflect this distribution. East of Tibet they are divergent, and west and southwest of Tibet convergent throughout the year. South and southeast of Tibet, the winds are convergent in winter and divergent in summer. These distributions derive from the combined mechanical–thermal effect of Tibet.

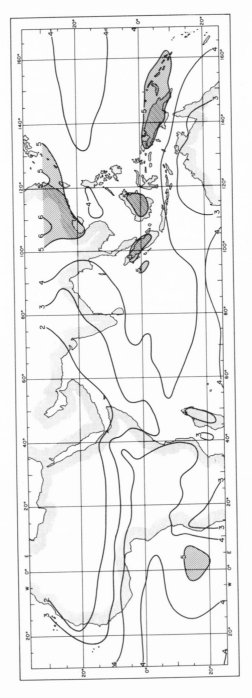

Fig. 5.1. Mean annual cloudiness (oktas) based on 3 yr of weather satellite photographs.

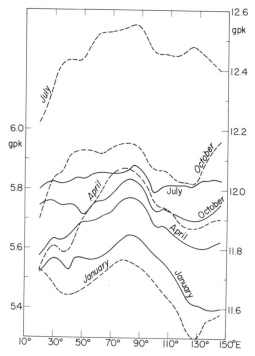

Fig. 5.2. Longitudinal variation of mean monthly pressure heights in geopotential kilometers at 500 mb (full lines) and 200 mb (dashed lines) in the zone 35–40N, for January, April, July, and October.

Figure 5.3 shows that upper-tropospheric pressure heights over southeastern Tibet (Lhasa) are greater than pressure heights to the west throughout the year, and pressure heights to the east during summer.

In the latitude strip including the northern rim of Tibet (Fig. 5.2), a sharp orographic ridge (Staff Members, 1958) persists between 80 and 85E throughout the year, unlike the situation farther south where the winter pressure maximum lies east of 120E. Figures 5.2 and 5.3 reveal that the largest winter meridional pressure gradients are likely between 30 and 35N, east of 90E.

5.1.1   Autumn, Winter, and Spring

Flohn (1968), confirming an earlier hypothesis (Staff Members, 1958), demonstrated from weather satellite pictures that central and southeastern Tibet are almost snow free during autumn, winter, and spring, and so act as a high-level radiational heat source. Pressure surfaces are raised there,

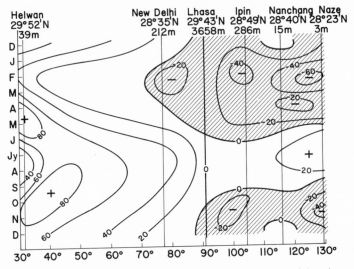

Fig. 5.3. Differences between mean monthly 300-mb pressure heights (geopotential meters) at Lhasa in southern Tibet and stations to east and west. Negative areas (Lhasa less) hatched. Patterns at 200 mb are similar, although complicated during winter by the presence of the tropopause. Temperatures recorded by Indian radiosondes are higher than those recorded by other radiosondes (Flohn, 1968; World Meteorological Organization, 1968). Accordingly, the New Delhi pressure heights have been adjusted.

diminishing the S–N temperature gradient and the strength of the subtropical jet to the south. Air subsides immediately east of the plateau and so is further warmed by compressional heating. Over central China it flows alongside very cold air which has swung around the northern edge of the massif (Plate I). Here the extremely large temperature gradient produces an exceptionally strong jet stream. With speeds increasing downstream eastward of 100E, upper-tropospheric divergence favors large-scale upward motion and increased cloudiness.

West of 100E, pressure heights smaller than over Lhasa, stemming from absence of a mid-tropospheric heat source, imply larger temperature gradients than to the south of Tibet. The consequently stronger, generally convergent upper-tropospheric westerlies over the Middle East favor sinking and decreasing cloudiness.

5.1.2 SUMMER

During summer, central and southeastern Tibet continue to act as a heat source, with condensation along the Himalayas powerfully aiding the effects of radiation. The subtropical ridge aloft overlies southern Tibet and compressional heating produced by mechanical subsidence is negligible. Consequently pressure surfaces at Lhasa are higher than to the east or west. Now the heating, instead of weakening a S–N temperature gradient as in winter, increases the N–S temperature gradient of the summer monsoon and produces a speed maximum in the upper-tropospheric easterlies south of India. The maximum possesses some jet attributes (Frost, 1953; Koteswaram, 1958). Upstream to the east of 70E, rising motion beneath divergent easterlies favors clouds and rain. Downstream to the west of 70E subsidence beneath convergent easterlies keeps skies clear.

Year-long subsidence persists over the great deserts of Southwest Asia and northern Africa, extending to the surface in winter anticyclones and to the lower troposphere in summer heat lows (Sections 3.1 and 3.2; Fig. 3.1).

5.2 Procedure Selection

In Chapter 2, I introduced the climatology of surface monsoon circulations in January and July. In Chapter 3, I discussed the range of monsoon components responsible for both fine and unsettled weather.

Now I turn to the interplay of climate and shorter period changes through the year.

5.2.1 PREVIOUS METHODS

The following possible procedures have been previously tried:

1. Separate the hemispheres and then divide the year into spring, summer, autumn, and winter. Simplification is achieved, however, at the expense of unified treatment of near-equatorial regions and of transequatorial circulation.

2. Separate meteorological elements into individual sections on temperature, pressure, wind, etc.; works by Braak (1921–1929), Lebedev and Sorochan (1967), and Rao and Ramamurti (1968) are examples. Graphs, histograms, and charts can be neatly fitted into the classification, but weather cannot.

3. Follow through the year region by region. This geographical approach is exemplified by Kendrew (1961), and Pédelaborde (1958). Reference searching is simplified and figures kept to manageable size, but the method may fail to reveal the underlying unity of monsoon phenomena.

5.2.2 METHOD SELECTED

I have used both the seasonal and geographical classifications. I first discuss the permanent *deserts* (Section 5.3) and then the *equatorial regions*

(Sections 5.4–5.6), proceeding from Africa to Indonesia and finishing with the Indian Ocean. The regions where winters are dry and summers are wet come next; the march of the seasons is followed separately over *northern Africa* (Section 5.7), *southern and Southeast Asia* (Section 5.8), and *southeastern Africa* and *northern Australia* (Section 5.9).

When necessary, I have introduced climatologically homogeneous sub-categories (Table 5.1) whose geographical boundaries approximate those delineated by Wissmann (1939) and Creutzberg (1950).

Plates I–IV will not be referred to specifically. However, they should be frequently consulted as providing the large-scale setting for the discussion.

Table 5.1
Climatological divisions of the monsoon area

	Classification	
	Wissmann[a]	Creutzberg[b]
1. Deserts		
Sahara	Desert	Dry
Northeastern Somalia	Desert	Dry
Arabia and West Arabian Sea	Desert	Dry
West Pakistan	Desert	Dry
2. Equatorial Regions		
Central Africa	Moist	Moist
Eastern Africa	Dry/weak dry period	Half moist/half dry
Indonesia and Malaysia	Moist	Moist
Indian Ocean	—	—
3. Africa		
Sudan (geographical definition)	Dry/dry period	Half moist/half dry
Southeastern Africa (Tanzania, Mozambique, and Madagascar)	Dry/dry period	Half moist/half dry
4. Southern Asia		
Himalayan foothills	Moist/dry summer	Moist
Central and East Arabian Sea	Dry period	—
India and East Pakistan	Dry period	Half dry/half moist
Bay of Bengal	Dry period	—
Burma and Thailand	Dry period	Moist/half moist
5. Eastern Asia and the China Seas		
China and the East China Sea	Moist	Moist
6. Australia		
Northern Territory and Queensland	Dry period	Half moist

[a] From Wissmann (1939).
[b] From Creutzberg (1950).

5.3 The Deserts

During April and October, the midmonths of the transition seasons, neither anticyclones (typical of the winter monsoon) nor cyclones (typical of the summer monsoon) dominate the surface circulation over the deserts. Diurnal temperature variation produces a significant circulation variation—relative low pressure prevails in the afternoon and relative high pressure at night. Over Arabia and the Arabian Sea diurnal variation sometimes shifts an anticyclonic cell to and fro between land and water.

During May and November the monsoon circulations become dominant throughout the day and intensify through the succeeding months of June and July, and December and January. Weakening begins in August and February with approach of the transition seasons.

Throughout the year, middle-tropospheric subsidence almost always prevents development of clouds deep enough to give rain (Fig. 5.4) (Hofmeyr, 1961).

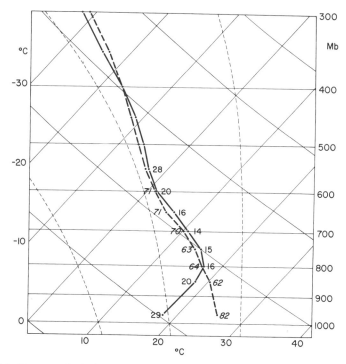

Fig. 5.4. Mean aerological soundings with relative humidity, for Khartoum (15°36′N, 32°33′E) at 0000 GMT February 1964 (full line) and at 0000 GMT August 1964 (dashed line).

Table 5.2
Monthly rainfalls at desert stations (1931 to 1960) in millimeters

		Jan.	Feb.	Mar.	Apr.	May	June	July	Aug.	Sept.	Oct.	Nov.	Dec.	Year
Colomb-Béchar 31°36'N, 2°10'W	Mean	7	9	13	8	3	2	0	4	7	14	13	10	90
	Heaviest	40	65	116	52	16	14	4	35	33	53	65	54	
Aoulef 27°00'N, 1°04'E	Mean	1	0.3	2	0.1	0.4	0.1	0	2	0	0.3	0.2	2	8
	Heaviest	25	3	14	1	6	4	0	20	6	17	20	16	
Ouallen 24°43'N, 1°15'E	Mean	2	0.4	0.6	1	0.5	1	1	2	2	1	0	3	14
	Heaviest	20	12	9	10	4	12	17	24	40	24	28	14	
Kidal 18°26'N, 1°21'E	Mean	0	1	0	1	4	8	38	51	28	1	0	0	132
	Heaviest	3	10	5	10	40	25	96	118	71	9	2	1	
Gao 16°16'N, 0°03'W	Mean	0	0	1	1	6	27	75	110	36	5	0	0	261
	Heaviest	7	4	7	7	62	73	141	228	81	24	1	4	

5.3.1 DESERT RAIN

Table 5.2 presents a north–south rainfall profile across the western Sahara from Colomb-Béchar to Gao. In the north transition months are wettest, whereas in the south summer is wettest.

Although the northwestern Sahara lies outside the monsoon area, cool-season situations which give significant rain there resemble cool-season rain situations over eastern Arabia and West Pakistan. Thus the results of careful studies made over the northwestern Sahara are probably valid for these other deserts as well.

5.3.1.1 *Northwestern Sahara.* Even in an exceptionally wet month, significant rain usually falls on only two or three days. Mayençon (1961) studied 33 occasions of heavy rain between 1950 and 1959 (Table 5.3). In midsummer,

Table 5.3

Occasions of heavy rain (> 30 mm day^{-1}) over the northwestern Sahara (1950 to 1959)[a]

Jan.	Feb.	Mar.	Apr.	May	June	July	Aug.	Sept.	Oct.	Nov.	Dec.	Year
2	1	4	2	1	0	0	2	6	8	3	4	33

[a] From Mayençon (1961).

locally heavy rain may fall from an isolated cumulonimbus in a rare moist incursion from south or southwest. Mayençon reported that in the other seasons, particularly during the September–November transition, if a large-amplitude trough or depression in the 500-mb polar westerlies approached or started to cross the Sahara while moist air of tropical maritime or equatorial origin covered the region, heavy rain sometimes ensued. Rising motion east of the upper trough lifted the moist air sufficiently to produce deep clouds and rain.

Jalu (1965) described similar sequences leading to surface cyclogenesis over the Sahara. He used weather satellite pictures in delineating cloud masses but did not report rainfall measurements. Cyclogenesis occurred on a front advancing ahead of a vigorous polar outbreak from the Atlantic or western Mediterranean. Upper-tropospheric divergence and incursion of relatively moist air south of the front preceded development.

Significantly, Jalu's analyses showed the heat trough well south of the polar front. Presumably, however, the trough circulation was involved in bringing the moist air northward.

5.3.1.2 *Southwestern Sahara.* In contrast to Colomb-Béchar where nine times as much rain fell during the wettest month on record as the average for March, the corresponding ratio for Kidal is only two. The southern desert enjoys a relatively reliable, though still meager, summer rainfall. Moist air in the heat trough may reach this far north. Occasionally, disturbances to the south (Section 5.7.4) destroy middle-tropospheric subsidence and thunderstorms develop.

During winter and the transition seasons drought is almost absolute. Troughs in the polar westerlies never extend so far south and subsidence is uninterrupted.

At Aoulef and Ouallen, where moist air seldom penetrates, significant rain neither falls in the transition seasons nor falls in summer.

5.3.1.3 *Other Deserts.* The meridional rainfall profile of the western Sahara typifies the eastern Sahara.

West Pakistan lies far enough north to be influenced by large-amplitude troughs in the polar westerlies during the cool season. The summer regime with the heat trough lying across the country resembles the summer regime of the southwestern Sahara. Thus at many desert stations rainfall varies bimodally with maxima in winter and summer.

Usually summer subtropical cyclones develop and remain almost stationary over the northeastern Arabian Sea (Section 3.5.4.1). With extreme rarity, one moves westward, or develops farther west. Then the southern coast of West Pakistan or eastern Arabia may be drenched. Between 17 and 23 July 1967, a subtropical cyclone deposited about ten times the normal July rainfall on eastern Arabia (Pedgley, 1970). Even more rarely, tropical cyclones may give rain to these regions during the transition seasons (Pedgley, 1969b).

5.3.1.4 *Northeastern Somalia.* This unusual equatorward protrusion of aridity results from an extremely dry summer (Flohn, 1964b). The cause lies in massive divergence arising from three interrelated mechanisms, which in combination are peculiar to the region:

1. The effect of the Great Rift heat trough on the southerly monsoon flow (Section 5.4.2.3).
2. The effect of stress differential on winds blowing parallel to a coastline with low pressure to the left (Section 2.2.2).
3. The effect of upwelling cold water in increasing the ocean–land pressure gradient and so increasing the wind speed downwind (Section 2.2.1).

5.3.2 Dust Storms

Any gust of wind raises dust in the desert, but great dust storms usually develop in the cool season. Then strong winds in polar outbreaks may sweep across the desert. Friction and convection raise clouds of dust several kilometers high, often obscuring the sun and reducing visibility to fog level. Winter dust storms have been well observed and described in the Sahara. Intense mesoscale dust storms, or "haboobs," summer phenomena, commonly occur around and to the south of Khartoum (Table 5.4).

5.3.2.1 *Winter Dust Storms.* A vigorous cold front moving rapidly southeastward across Morocco heralds the onset of the harmattan and possibly a severe dust storm. When the strong winds hit, the desert dust rises. However, the front, once over the desert, is hard to follow. Hamilton and Archbold (1945) and Johnson (1964c) stated that freshening northeast winds often begin to blow dust well ahead of where the front might reasonably have been expected to be.

This puzzling observation is dramatically duplicated when a vigorous surge of the winter monsoon reaches the South China Sea. As described in more detail in Section 5.8.2.3, the cold front at the leading edge of the surge is quickly lost as northerlies freshen almost simultaneously to 10N and beyond.

Jalu and Dettwiller (1965) described an exceptionally severe dust storm which on 27 March 1963 reached 10N and raised thick dust over an area of 300,000 km^2 in the central Sahara. Both at the surface and aloft the situation closely resembled the East Asian cold surge described in Sections 3.1.3.1 and 5.8.2.3. Behind the rather sharply defined edge of the advancing dust cloud, surface temperatures were more than 10C lower than in the clear air ahead. Jalu and Dettwiller ascribed the difference to interception of solar radiation by the thick dust. Their conclusion that the dust served to accentuate the cold front along its leading edge is not necessarily valid. If events had likely followed the sequence described in the previous paragraph, the front would have long since disappeared. Instead the dust would have been continually creating its own accompanying "front." The large temperature gradient across the leading edge of the dust cloud would also have produced a strong low-level jet similar to the jet discovered by Bunker (1967) above the cold Somali Current (Section 2.2.4). Probably then, once a threshold of radiational opacity is crossed, feedback may well intensify a dust storm.

5.3.2.2 *Summer Dust Storms.* Table 5.4 shows that winter dust storms moving down from the north affect Khartoum, but that summer dust storms, or haboobs, which generally move in from the south, are much more common. According to Sutton (1925) they are convective phenomena; one in three was followed by a thunderstorm and two in three by rain. The storm fronts were usually between 20 and 30 km long and affected a locality for about 3 hr. Haboobs are generally of mesoscale size, most frequently occur between 1600 and 1800 hr, and can seldom be traced between neighboring stations.

Table 5.4

Frequency of dust storms at Khartoum (15°36'N, 32°33'E) (8 yr)[a]

		Jan.	Feb.	Mar.	Apr.	May	June	July	Aug.	Sept.	Oct.	Nov.	Dec.	Year
Total		5	7	11	13	21	48	37	23	25	5	0	1	196
Direction of approach[b]	NE–NW	5	7	8	4	3	5	1	1	1	2	0	1	38
	SE–SW				4	12	26	16	11	15	2	0		86

[a] From Sutton (1925).
[b] Direction of approach could not be determined for all storms.

They originate in complex interactions between synoptic disturbances and local orographic and heating gradients.

Rain accompanying the storm lays the dust and reduces the chances of a rapid sequence of storms. Sutton thought that the frequency decrease between June and August at Khartoum (Table 5.4) might be partially due to sprouting of a dust-inhibiting ground cover. Similar dust storms are common during summer near the borders of other deserts (Powell and Pedgley, 1969).

At Agra (27°09′N, 77°58′E) 80% of all dust storms occur in May and June (one day in four) prior to summer monsoon rain which is as plentiful as at Khartoum. The storms are also mesoscale instability phenomena; 64% occur between 1500 and 2000 hr and only 25% between 0400 and 1100 hr. In spring they may be triggered by western disturbances (Sreenivasaiah and Sur, 1939).

5.4   Equatorial Africa

Between 5N and 5S the annual variation of average monthly rainfall tends to be bimodal (Fig. 2.6), with a maximum in each transition season (Fig. 5.5.)

West of 30E and the escarpments of the Great Rift, the monsoon regime is weak and confined north of the equator and east of Gabon and Cameroon by the relatively cold waters south of the Gulf of Guinea (Section 2.3.1.4).

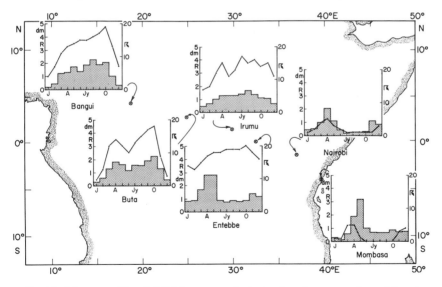

Fig. 5.5. Mean monthly rainfall (histograms; decimeters), and mean numbers of thunderstorm days (full lines), across equatorial Africa.

Surface winds from a westerly quarter predominate and moist air may penetrate to the interior in any month.   East of 30E, monsoons dominate on both
sides of the equator. Surface winds from south of east prevail between April
and October and from north of east for the rest of the year.

Westerlies west of 30E and easterlies east of 30E require low pressure along
30E, and, in the Great Rift, persistent high temperatures maintain a heat
trough (Weickmann, 1963).

Rainfall variation possesses several periods (Fig. 2.8); apart from the
diurnal, the synoptic (2–5 days) and another, ranging from 2 to 5 weeks, seem
to predominate. In higher monsoon latitudes, the former can often be related
to well-defined disturbances, but thus far this has not proved possible near the
equator (Section 3.9.1). The latter, which is discussed in Chapter 7, is more
obscure; perhaps it stems from very large-scale changes within the monsoon
area or between the monsoon area and regions beyond.

Although moisture is essential for clouds and rain, there is always sufficient
available in the lower layers over equatorial regions (Johnson and Mörth,
1960). Horizontal moisture gradients are small, and so vertical motion, not
advection, determines the depth of the moist layer (Godske et al., 1957), an
opinion confirmed by a numerical modeling experiment (Krishnamurti, 1969).

5.4.1   EQUATORIAL AFRICA WEST OF 30E

5.4.1.1   *Northern Hemisphere Summer and Transition Seasons.* During these
months the air is moist in the lower layers and unstable enough to favor the
triggering of thunderstorms which occur on half of the days. In the transition
seasons, southern parts of squall lines occasionally move westward across the
region (Figs. 3.41 and 3.42).

Surface westerlies prevail between the heat trough to the north, a near-
permanent heat low to the south (Section 2.3.2.4), and the Great Rift heat
trough. When the westerlies strengthen and deepen to 2500–3000 m (Jeandidier and Rainteau, 1957; Johnson, 1964c), synoptic-scale lifting causes
deep nimbostratus to form. Lapse rates approach the saturated adiabatic,
thunderstorms are rare, and a *rains* regime prevails (Section 4.1.1).

This active period, according to Tschirhart (1959), is distinguished by
quasi-stationary zonal perturbations (Section 3.9.1) typically consisting of a
somewhat fragmented east–west band of weather 2000–3000 km long and
200–300 km wide. Thick chaotic middle clouds first develop and within them
cumulonimbus form. The perturbations, which last about 24 hr, move slightly
and irregularly, exhibiting little diurnal variation or thunderstorm activity. In
fact their presence inhibits squall development. On an average, one to two perturbations develop per month, but sometimes a month may go by without one.

Surface equatorial westerlies, associated with unsettled weather, are classed as a "bridge" situation by Johnson and Mörth (1960) (see Flohn, 1960c) (Fig. 3.33). Near the equator, they considered that the term $2\Omega \cos \phi \, dx/dt$ in the vertical motion equation $dw/dt = -(1/\rho)(\partial p/\partial z) + 2\Omega \cos \phi \, (dx/dt) - g$ could produce vertical accelerations which upset the hydrostatic balance between $(1/\rho)(\partial p/\partial z)$ and $-g$. Although compensation would damp the acceleration, it might still appreciably affect the weather. Westerlies would favor rising and easterlies sinking motion.

Gordon and Taylor (1970) applied their trajectory equations (Section 3.10.1.2) to zonal equatorial pressure gradients. They determined that, in a west–east gradient (westerlies), the air converged along the equator and, in an east–west gradient (easterlies), it diverged. In explanation it is unnecessary to adduce changes in $2\Omega \cos \phi \, dx/dt$; the meridional gradient of the coriolis parameter ($2\Omega \sin \phi$) is sufficient to account for convergent westerlies and divergent easterlies. The methods differ, but the results agree. All would now be well if nature obliged. She does with equatorial westerlies; but, as Section 5.4.2.1 shows, very unsettled weather can occur with equatorial easterlies.

When the westerlies slacken and diminish in depth to below 2000 m, dryness and increased instability aloft encourage thunderstorm development, particularly along the edge of the eastern escarpment. Falls may be locally heavy but are usually scattered.

Jeandidier and Rainteau (1957) and Johnson (1964a) described *eastward-*moving areas of bad weather. Lack of upper-air data precludes definite conclusions. However, analogous sequences often occur over the central Pacific. Apparently some eastward-moving troughs in the polar westerlies are of such intensity and amplitude that upward motion ahead of them may worsen near-equatorial weather. Although the troughs are stronger in higher latitudes, the air is too desiccated there for lifting to produce significant clouds. Thus meridional continuity can seldom be unequivocally established. According to Thompson (1965), Southern Hemisphere polar troughs affect the weather more often than Northern Hemisphere polar troughs, penetrating farthest north in April.

5.4.1.2 *Northern Hemisphere Winter.* Surface winds are often variable. The Atlantic westerlies seldom penetrate beyond the coastal ranges while occasional northerlies bring in very dry air. Consequently, little rain falls. Large-amplitude polar troughs rarely induce a deterioration.

5.4.2 EQUATORIAL AFRICA EAST OF 30E

Eastern equatorial Africa is considerably drier than western equatorial Africa, probably because westerlies predominate in the west and easterlies in the east (Section 5.4.1.1). Thunderstorms are very common near the western highlands, but they and squall lines are rare to the east, being almost non-existent during the monsoons (Fig. 5.5).

5.4.2.1   *The Transition Seasons.* In contrast to the west, where cold South Atlantic coastal waters permanently distort the circulation, a circulation typical of the tropical oceans prevails in the east. Trade winds, on the peripheries of anticyclones to north and south, apparently converge in a broad low-pressure zone along the equator—according to Johnson and Mörth (1960), a "duct" situation (Fig. 3.33). However, heat troughs occasionally appear over the drier regions between 5 and 10° of latitude. If, as suggested in Section 5.4.1.1, equatorial easterlies are divergent, why should rains be heaviest in the transition seasons? Although under a purely zonal pressure gradient, flow diverges; addition of a slight equatorially directed pressure-gradient could produce convergence. The changes may be subtle. Often *weather* supplies the first evidence—too late to help the forecaster.

The convergence and bad weather wax and wane over periods of a few days (Johnson, 1962), apparently responding to intensity fluctuations in the anticyclones, which in turn may be influenced by events in higher latitudes.

In earlier times the near-equatorial easterlies were said to coincide with the intertropical convergence zone, and meteorologists ascribed the annual rainfall variation to movement of the zone toward and away from the equator (Henderson, 1949). This concept may have climatological relevance but cannot be applied to day-to-day forecasting. Convergence zones develop and dissipate as new zones appear hundreds of kilometers away (Johnson, 1962; Thompson, 1965) (Section 3.8.1).

5.4.2.2   *Development and Decay of the Monsoons.* As the solstices approach, the near-equatorial low-pressure zone of the transition seasons does not move steadily toward the summer pole but, rather, weakens and disappears as the weak heat trough in the summer hemisphere intensifies over the deserts and moves unsteadily poleward (Section 5.7) (Thompson, 1965). In the winter hemisphere, pressures rise over the deserts. Thus, rather abruptly, in May and November, the easterlies are replaced by transequatorial flow. The "drift" (Section 3.10) now predominates over the "duct" (Fig. 3.33), although for a few weeks weather is more than usually chaotic.

The sequence reverses near the equinoxes. In March and September, lowest pressure is once more established near the equator and the "duct" predominates over the "drift."

5.4.2.3   *The Monsoons.* Mean-flow charts prepared by Findlater (1968) confirm that low-level divergence prevails during both monsoons (Glover *et al.*, 1954). The Great Rift heat trough seems to be the cause. In January, the northeast monsoon divides, with one branch flowing toward the heat low over Mozambique and the other flowing westward toward the Great Rift (Cochemé, 1960). In July, a minor branch of the southerly monsoon diverges toward the Great Rift while the main stream continues northward.

The rainfall profiles shown in Fig. 5.6 are readily related to the divergence profiles of Figs. 3.34 and 3.37. The differences—relative dryness in the north in January and in the south in July as well as the overall dryness in both months—might arise from the effect of the Great Rift heat trough. Consonantly, at 200 mb slightly convergent winds blow toward the winter hemisphere.

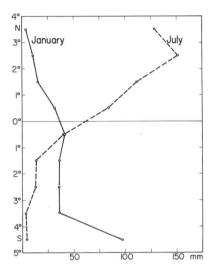

Fig. 5.6. Meridional profiles of median rainfall for January and July over eastern Africa.

5.4.2.4 *Diurnal Variation of Rainfall.* Twenty-four-hour rainfall totals may reflect presence or absence of synoptic-scale disturbances (Johnson, 1962). But, as in other near-equatorial regions (Section 4.1.3), local topography apparently determines at what hours the rain falls (Thompson, 1957b, 1968).

A strange regime prevails along the southern slopes of Mount Kilimanjaro (3°04′S, 37°22′E) during the southeast monsoon. According to Thompson (1957b), thick altostratus or nimbostratus forms at night, giving widespread *rains*. Sometimes violent thunderstorms develop from embedded cumulonimbus. By afternoon, only scattered convective clouds remain. Ridges extending southeastward from the mountain mass funnel the air as it ascends the main slope. Here, perhaps orography duplicates the synoptic-scale convergence which elsewhere produces monsoon *rains*.

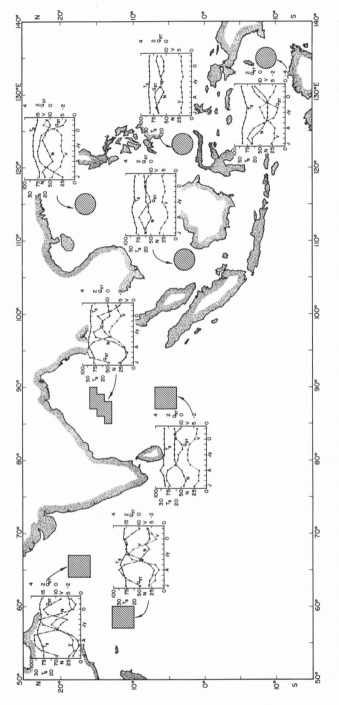

Fig. 5.7. Annual variation of meteorological elements over selected ocean areas. Wind speed (V) in meters per second, cloudiness (N) in percent, sea-surface temperature (T_s) in degrees centigrade, and net heat balance at the sea surface (Q_{ST}) in hectolangleys per day (Ramage, 1969b; Wyrtki, 1957).

5.5 Indonesia and Malaysia

On reading Braak's authoritative treatise on the climate of Indonesia (Braak, 1921–1929), one is almost overwhelmed by the myriad subclimates into which the monsoons are distorted by the great mountainous islands of the "maritime continent." Within the envelope of the archipelago no place is free from these distortions. However, strung through the enclosed seas are several small islands, beyond the immediate influence of large-island diurnal cycles, on which meteorological measurements have been made. Discussing the march of the seasons on these islands and over their surrounding seas (Fig. 5.7) first, will provide a basis for understanding the complex role of the large islands in Indonesian weather.

5.5.1 THE "UNDISTORTED" MARCH OF THE SEASONS

5.5.1.1 *Data.* Figure 5.8 shows mean monthly rainfalls at eight and mean monthly resultant surface winds at four stations on small islands.

Wyrtki (1956) evaluated the rainfall distribution over the Indonesian seas by first constructing a chart of average annual rainfall for the region. He utilized stations on small islands and assumed that most of the difference between rainfall over the large islands and rainfall over the sea is concentrated

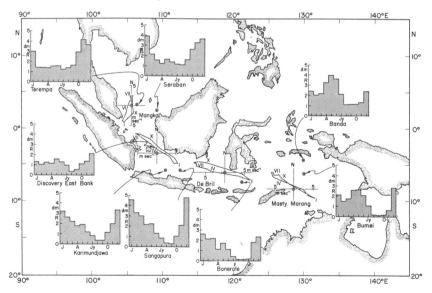

Fig. 5.8. Mean monthly rainfalls (decimeters) and mean monthly resultant surface winds (m sec^{-1}) at stations on small islands (from Braak, 1921–1929; Preedy, 1966).

near the coast. He then divided the seas into 12 geographically and climatologically homogeneous subregions (Fig. 5.9). Within each subregion observations from a number of suitable island and coastal stations were considered to determine the average annual variation of rainfall in the subregion. Finally, he obtained the "true" annual subregional variation by multiplying the average for each month by the ratio of the annual subregional average (determined from his chart of average annual rainfall) to the subregional average (determined directly from the observations) (Table 5.5).

Figure 5.10 presents monthly weather at Djakarta during 1963. Temporal incompatibility of surface and upper-air data is not uncommon in the tropics. All the same, 1963 could be a lucky choice since at Djakarta the wet season was much wetter and the dry season was much drier than normal. Thus the variation during this year probably resembles the average annual variation, but is more accentuated.

5.5.1.2 *Southern Hemisphere Winter, the Southeast Monsoon.* Rainfall is meager, being least in the southeast and most in the northwest. The cause is subsiding outflow from the Australian anticyclone. In the southeastern half

Fig. 5.9. Subregions selected for the calculation of the annual variation of rainfall. Numbers refer to Table 5.5 (Wyrtki, 1956).

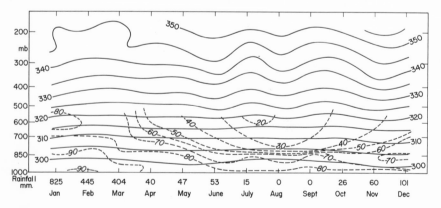

Fig. 5.10. Monthly means for 1963 at Djakarta (6°11′S, 106°50′E). Isentropes (°K) (full lines), isohumes (percent) (dashed lines), rainfall (monthly totals).

Table 5.5

Average amounts of rainfall (mm) over different parts of the Indonesian waters[a]

Subregion	Jan.	Feb.	Mar.	Apr.	May	June	July	Aug.	Sept.	Oct.	Nov.	Dec.	Year
1. Malacca Strait	163	95	124	149	151	122	119	152	187	224	230	234	1950
2. South China Sea	251	141	153	153	160	145	124	121	164	233	270	295	2210
3. Java Sea	244	185	171	140	150	107	79	49	39	81	128	217	1590
4. Flores Sea	216	152	154	148	139	108	54	20	18	41	100	190	1340
5. Banda Sea	183	140	156	189	203	179	118	58	32	42	70	170	1540
6. Arafura Sea	238	198	205	220	229	143	91	48	44	53	94	187	1750
7. Celebes Sea	188	143	170	166	175	174	148	136	134	145	183	188	1950
8. North of New Guinea	214	191	206	187	182	182	149	138	132	114	140	155	1990
9. Molucca Sea	148	134	163	188	218	216	171	110	104	90	112	136	1790
10. Savu Sea	217	186	155	65	32	26	19	8	9	29	75	154	975
11. South of Java	226	200	192	148	113	99	67	40	46	133	215	246	1725
12. Southwest of Sumatra[b]	273	241	255	263	210	182	194	239	288	366	383	336	3230

[a] From Wyrtki (1956).
[b] See Section 5.6.2.

of the archipelago typical tradewind conditions prevail (Braak, 1921–1929; Depperman, 1941). Beneath a sharp inversion at about 2000 m only scattered cumulus or stratocumulus develop. Haze, originating from both the ocean (salt particles) and continent (dust), may seriously reduce visibility (van Bemmelen, 1905). From May through October 1963 dry subsiding air covered Djakarta above 850 mb. Winds diminish and back to south at the equator, where they join the summer monsoon of Southeast Asia. Subsidence diminishes downstream; remnants of the dry layer can rarely be detected at Singapore. At 200 mb, mean resultant winds are weakly convergent over central Indonesia and weakly divergent to the west.

5.5.1.3 *The October–November Transition.* Typhoons often form in the southward-moving Northern Hemisphere near-equatorial trough. Circulation around the weakening Australian anticyclone hinders development of a Southern Hemisphere near-equatorial trough. Consequently, a northward-directed transequatorial "drift" prevails (Section 3.10, Fig. 3.33). Therefore average rainfall is greater north of the equator than south of the equator; heavy falls occur in southwesterly surges accompanying typhoon development to the north. In 1963 the middle troposphere over Djakarta had become moister.

5.5.1.4 *Northern Hemisphere Winter, the Northwest Monsoon.* As the pressure gradient between the Siberian high and Australian heat low builds up, the northern hemisphere near-equatorial trough weakens; flow of air across the equator begins by December and reaches a peak in January. Stations south of the equator experience their heaviest rainfall in the anticyclonically curving "drift" (Section 3.10, Fig. 3.33), while amounts begin to decrease north of the equator. At Djakarta in 1963 air was moist to above 500 mb, although the lower troposphere was slightly more stable than in the winter. Extreme instability in the high troposphere is alluded to in Section 7.2.1.

At 200 mb mean resultant winds are generally divergent.

5.5.1.5 *The March–May Transition.* As the Siberian high and Australian heat low weaken, near-equatorial troughs are found in both hemispheres. Over the South China Sea, near absence of typhoons leads, near the equator, to reduced cloudiness, light winds, and maximum heat gain to the ocean surface. A lingering effect of high pressure over China is revealed by the east–northeast wind at Mangkai, whereas at Discovery East Bank the resultant wind is west of south.

In subregions 5, 6, and 9 (Table 5.5) most rain rather unaccountably falls in the period from April to June. According to Otani (1954) a pressure trough lies across this region in winter and associated convergence could lead to increased rain. What causes the trough?

The climates of the small islands reflect the interacting influences of Asia and Australia. The Asian continental circulations are intense but distant, while the Australian continental circulations are weak but close. Consequently Southern Hemisphere winter subsidence inhibits rain whereas Northern Hemisphere winter subsidence is effective only north of the archipelago (Fig. 5.46).

5.5.2 INDONESIAN WEATHER AND THE ROLE OF LARGE ISLANDS

The large islands and Malaya massively modify near-equatorial weather on many time and space scales. Rather than attempt a detailed description, I shall provide examples of orographic, diurnal, and large-scale effects. The reader should keep in mind that ascending motion is generally characteristic of the northwest monsoon and subsidence of the southeast monsoon (Depperman, 1941).

5.5.2.1 *Orographic and Diurnal Effects.* Great mountain ranges such as those of Sumatra and New Guinea enhance rainfall to windward. When the monsoon is vigorous, *rains* prevail. Diurnal variation is slight or has even a nocturnal maximum (Section 4.1.3.2), typically along the north coast of Java in January and to windward of the western New Guinea range in July.

During the Northern Hemisphere summer, southwesterlies may become sufficiently strong over northern Sumatra to produce a föhn effect east of the main range (Braak, 1921–1929). Besides being dried, the air is destabilized by the descent and occasionally severe late afternoon thunderstorms develop (Section 4.1.3.2). Thunderstorm squalls and föhn gusts then combine to wreak considerable local damage (Zain, 1969).

However, at the height of the monsoons, winds are rarely strong enough to produce severe föhn desiccation leeward of the ridges. There, on most days, mountain and sea-breeze circulations lead to development of vigorous afternoon showers (Fig. 4.6).

Bryson and Kuhn (1951) used New Guinea to illustrate the effect of stress differential on rainfall (Section 2.2.2). In the southeast monsoon, flow parallels the main range and long sections of the coastline. Along the northern coast, convergence induced by stress differential enhances rain while along the southern coast divergence induced by stress differential inhibits rain (compare subregions 8 and 6 in Table 5.5).

5.5.2.2 "*Sumatras.*" Between April and October southwesterlies prevail over the Strait of Malacca. Along both shores of the Strait nocturnal land breezes are common and converge with undisturbed air 10–20 km offshore.

The convergence is sometimes vigorous enough to erect a line of cumulo-nimbus with squalls and thunderstorms—the "Sumatra." Sumatras may develop off any of the large islands (Braak, 1921–1929). A concave-shaped coastline and mountain range cause additional convergence *within* a land breeze and so accentuate convection (Neumann, 1951). Consequently, Sumatras often develop off the north coast of Sumatra between Kutaradja (5°34′N, 95°20′E) and Langsa (4°28′N, 97°58′E) (Braak, 1921–1929), where they tend to remain offshore, and off southwestern Malaya between Port Swettenham (3°00′N, 101°25′E) and Singapore (Benest, 1923; Watts, 1955). The southwest monsoon helps generate Sumatras by reinforcing the land breeze leeward of the Sumatra ranges (Braak, 1921–1929). Sumatras often move up to 50 km inland over Malaya, perhaps because cold downdrafts from the cumulonimbus first sweep eastward with the current at 2000 m then dip to the surface where lifting along a pseudofront forms another series of cumulonimbus columns (Watts, 1955). Certainly origins and movements of Sumatras and West African squall lines (Section 3.11.2) differ.

As in other equatorial regions, diurnal variations in rainfall differ from place to place and from season to season (Section 4.1.3). During the transitions, when general flow is light, diurnal variations in local winds largely determine diurnal variation of rainfall. Then, afternoon thundershowers usually predominate (Braak, 1921–1929; Preedy, 1966).

5.5.3 WEATHER CHANGES

According to Braak (1921–1929), tropical cyclones rarely affect the archipelago and then only between late March and early May in the extreme southeast and between October and December south of the Philippines. As the cyclones move westward, they induce heavy *rains*, but seldom destructive winds.

Otherwise, variations of rainfall occur with periods varying, according to Braak, "from some days to half a month and sometimes longer." From a single vantage point, the sequences might be thought to result from passing depressions, but pressure does not change appropriately and according to Braak "a progressive movement cannot be detected in these rains." The areas covered by the *rains* are much smaller than in higher latitudes—sometimes not more than half of Java.

Johnson's descriptions of East African weather systems (Section 3.9.1), strikingly echo Braak's descriptions of Indonesian weather systems, and weather satellite pictures confirm their size and relative immobility (Fig. 3.30).

Watts (1955) related weather changes to movement of "air stream boundaries" but the boundaries, lying within a single air mass, cannot long be followed nor can their apparent movements be extrapolated with any degree of confidence.

Schmidt (1949) reported that small east–west-oriented disturbances cause unsettled weather over Java after moving northward across the Timor Sea. His attempt at a mathematical formulation failed.

5.6 Indian Ocean

Under the influence of Africa in the west and Australia and Indonesia in the east, the monsoons extend well across the equator. In the center, they are confined north of the equator.*

5.6.1 WESTERN INDIAN OCEAN

The East African regime (Section 5.4.2) extends eastward to 50E, with most rain in the transitions. The quite different regime of the Seychelles (Table 5.6) has puzzled meteorologists (Riehl, 1954). The January maximum coincides with a persistent "drift" (Section 3.10, Fig. 3.33) in which flow from the Northern Hemisphere curves anticyclonically and converges over the islands beneath divergence at 200 mb. Rare thunderstorms and negligible diurnal variation of rainfall (Thompson, 1957b) are features of a *rains* regime. Conversely, in July, the islands lie on the winter side of a "drift," where equatorward flow diverges (Section 3.10.1.3), beneath convergence at 200 mb.

Variable winds in April and November, a diminished rainfall decrease between April and May, and an enhanced increase between October and November confirm that in the transitions near-equatorial troughs affect the Seychelles. Soon, however, changes over the continents to north and south dominate the circulation and initiate transequatorial flow. Perhaps a study of five-day means would uncover minor transition rainfall peaks.

Unaffected by the Great Rift heat trough (Section 5.4.2.3), Seychelles rainfall resembles, in its annual variation, the rainfall of small islands in the Java Sea (Fig. 5.8) and differs sharply from the East African regime (Fig. 5.5).

5.6.2 EASTERN INDIAN OCEAN

Here, on the edge of the monsoons, doldrum effects are pronounced. On the island chain west of Sumatra (Fig. 5.11), October marks the rainfall

---

* At Gan (0°41′S, 73°09′E) lower-tropospheric winds from a westerly quarter prevail throughout the year.

Table 5.6

Climate of Mahé (4°37'S, 55°27'E) in the Seychelles[a]

	Jan.	Feb.	Mar.	Apr.	May	June	July	Aug.	Sept.	Oct.	Nov.	Dec.	Year
Rainfall (mm)	386	267	234	183	170	101	84	69	129	155	231	340	2349
Raindays	15	10	11	10	9	9	8	7	8	9	12	15	123
Thunder days	3	3	3	2	2	1	1	1	1	1	1	3	18
Surface wind direction	NW	NW	NW	Var	SE	SE	SE	SE	SE	SE	Var	NW	
Speed (m sec⁻¹)	3	3.5	2.5	2.5	3.5	4.5	6	6	5	3.5	3	2.5	

[a] From Preedy (1966).

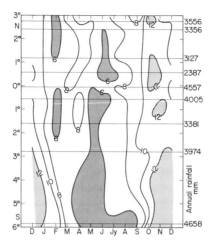

Fig. 5.11. Mean monthly rainfalls expressed as percentages of the mean annual rainfall for each of nine island stations off the western coast of Sumatra (Braak, 1921–1929; Preedy, 1966).

maximum in the north, November or December in the south. The driest months are usually May or June. Cloudy skies and frequently heavy rains accompany strong west and northwest winds (Braak, 1921–1929), apparently a "bridge" situation (Fig. 3.33, Section 5.4.1.1). The sea surface gains the least heat in the Northern Hemisphere winter because winds are then strongest and evaporation greatest.

Resemblance to the annual variation of rainfall on small islands in the Indonesian seas (Fig. 5.8) is blurred by the fact that off Sumatra the driest month is 45% as wet as the wettest month, whereas in the Indonesian seas the ratio is only 17%.

5.7 North Africa

Annual variations in this region are more regular than anywhere else in the monsoon area. From one day to the next, the east–west-oriented heat trough may move discontinuously. Through this "noise," the trough advances slowly northward from about 7N in January and February to about 22N in August (Fig. 5.12). That it then returns southward much more quickly (Bhalotra, 1963; Garnier, 1967) is typical of the monsoon area and stems from the fact that the normal meridional temperature gradient is *reversed* by the summer monsoon, but *increased* by the winter monsoon.

Moisture is in better supply in the west of the region than in the east and so cloudiness and rainfall diminish eastward, except where orographically

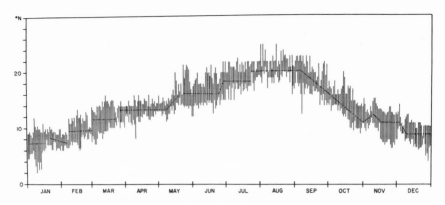

Fig. 5.12. The range of location of the heat trough line at 0600 GMT, along 3E, 1956–1960 (Garnier, 1967).

enhanced by the Ethiopian highlands. Nevertheless, on climatological charts (Thompson, 1965), isopleths tend to parallel the heat trough and to move in unison with it. The various weather regimes are also disposed in zones, the widths of which may fluctuate widely (Hamilton and Archbold, 1945; Walker, 1960; Flohn, 1960b; Garnier, 1967).

5.7.1   WEATHER ZONES

In what follows I use Garnier's definitions and relate them to the features discussed in Chapter 3.

5.7.1.1   *Zones A and B.* Zone A lying immediately to the north and Zone B lying immediately to the south of the heat-trough line encompass the fine weather associated with midtropospheric subsidence above the heat trough (Section 3.2.1). Zone A is dry and dusty; Zone B, extending 250–300 km south of the trough line, is moist beneath the inversion. Early morning mist or scattered afternoon cumulus may form.

Fig. 5.13. Kinematic analysis and isobaric analysis in perspective for 1200 GMT, 23 August 1967. From top: (1) 3 km, with stations reporting 2- and 4-km winds denoted by triangles; (2) 0.6 km, with stations reporting surface winds denoted by triangles; (3) isobars labeled in millibars and tenths above 1000; (4) clouds photographed from ESSA 5 between 1308 and 1655 GMT, 23 August 1967 (Orbit Nos. 1585–1587). Longitudinal positions of the upper wave axes are designated by arrows at the tops of (1) and (4) (Carlson, 1969a).

The diagram shows two westward-moving middle tropospheric cyclones. Typical *rains* clouds extend southwest from the eastern cyclone. Stress-differential induced divergence (Section 2.2.2) may account for lack of clouds west and southwest of the western cyclone. Along the heat trough, near 20N, skies are clear.

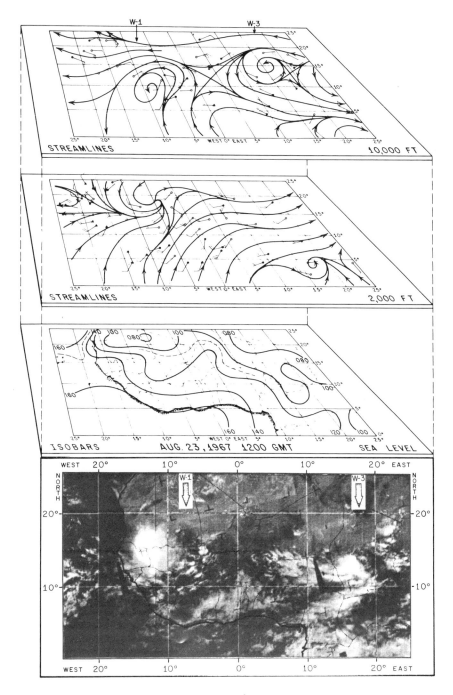

Fig. 5.13

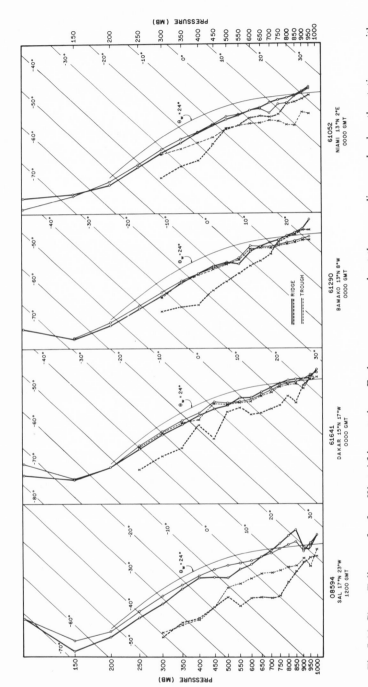

Fig. 5.14. Skew-*T* diagram for four West African stations. Each curve averages about three soundings made when the station was either close to the axis (light lines) of one of four major waves which crossed the region between 21 August and 5 September, 1967 or near the intervening ridge lines (heavy lines). Temperatures are represented by full lines and dew points by dashed lines. The wet-bulb potential temperature curve, $\theta_w = 24C$, is also drawn (Carlson, 1969a).

As with subtropical cyclones (Section 3.5) the troughs are colder than the ridges below about 600 mb and warmer above (Fig. 3.10).

5.7.1.2 *Zone C.* Zone C extends 760–800 km south of Zone B. The air is usually moist and unstable and *showers* prevail (Section 4.1.1).

West of 25E, squall lines, an example of which is presented in Section 3.11.2.1, and disturbances resembling subtropical cyclones (Section 3.5, Figs. 5.13 and 5.14) move westward across the zone accompanied by severe thunderstorms.* Thunderstorms may also break out for no apparent reason. Most rain falls in the early evening (Delormé, 1963).

East of 25E, moving systems are rare (Pedgley, 1969a) although some thundery disturbances drift a short distance westward from the Ethiopian highlands before drying out. Pedgley identified two diurnal rainfall regimes. Regions more than 700 km west of the Ethiopian highlands experience a late-afternoon or early-evening maximum. To the east during the day, a "hot plume" of air spreads from the Ethiopian highlands across the adjacent plains and, by raising temperatures in the middle troposphere, suppresses the normal afternoon maximum. Diurnal variation of rainfall is thereby reduced and a weak nocturnal maximum often appears, perhaps stemming from the drifting thunderstorms mentioned above.

5.7.1.3 *Zone D.* In Zone D, which extends up to 300 km south of Zone C, *rains* prevail when the southwesterlies deepen (Section 5.4.1.1). Even when conditions are more settled, skies often remain overcast with low stratus, diurnal variation is slight (Delormé, 1963), and rainfall on the average is less than in Zone C.

Figures 5.15 and 5.16 illustrate the meridional progression of weather zones and some of their characteristics.

The annual variation of rainfall (Fig. 5.15) has a double peak below 7N and a single peak above, indicating that Zone D seldom moves above 7N. Conversely, since below 3N it rains in every month, Zone A never reaches so far south. Between 9 and 15N, August is the wettest month, suggesting that the heaviest rainfall, associated with Zone C, seldom moves north of 10N. Figure 5.16 should be compared with Figure 4.2.

5.7.2 WINTER

Fort Lamy/Maiduguri lies well north of the heat trough in Zone A. Lagos and Douala/Calabar are predominantly in Zone B.

West of 0°, northeast and southeast tradewinds converge just north of the

---

* The squalls analyzed in Fig. 5.16 may accompany either squall lines or cyclonic disturbances. The reader should not confuse the term "disturbance," as used here, with "disturbance line" which to some authors is synonymous with "squall line."

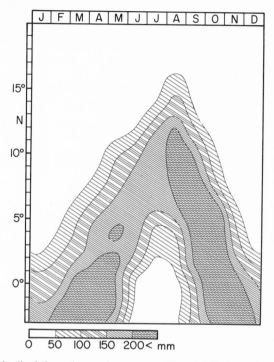

Fig. 5.15. Latitudinal dependence of mean monthly rainfall along the meridional strip between 10 and 15E (Hendl, 1963).

equator in a band of heavy rain. Release of latent heat results, at 200 mb, in a ridge and westerlies along the equator.

East of 0°, the ridge trends north. The net effect of this confluence between southwesterlies and westerlies (Namias and Clapp, 1949) is massive convergence leading to mid-tropospheric subsidence and drought over Africa north of 5N. A close parallel is observed over the equatorial eastern Pacific in January—very heavy rains just north of the equator near 120W and drought over Panama and Costa Rica. Here, too, cold water upwells to the south.

5.7.3 SPRING TRANSITION

The heat trough usually moves north of Fort Lamy/Maiduguri early in April (Fig. 5.12). By then the sun's heat over western Africa overcomes latent heat release near the equator and the 200-mb ridge now lies along about 8N.

Fig. 5.16. Squalls at three groups of stations in and near Nigeria. Average frequency of squalls (Hamilton and Archbold, 1945) (full lines), average dew point depression at 850 mb (dotted lines), and average difference in virtual temperature between 850 and 300 mb (dashed lines).

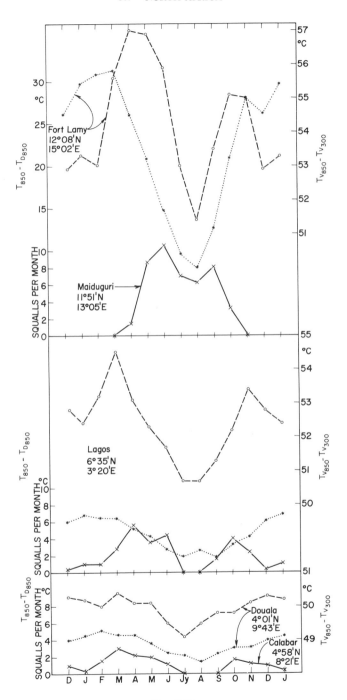

Fig. 5.16

The winter convergence at this level has disappeared and instability reaches a maximum. At Fort Lamy/Maiduguri instability combines with increasing lower-tropospheric moisture content to favor squalls, which become most frequent in June when the station is usually in Zone C. At Khartoum, most haboobs occur in June (Table 5.4) (Sutton, 1925).

At the southern stations, instability is greatest in March, when Zones B and C alternate, and then decreases as Zone C predominates. Most squalls occur in April at Lagos and a month earlier at Douala/Calabar.

5.7.4   SUMMER

The heat trough is near 20N, upper-tropospheric easterlies cover the region. The influence of Himalaya–Tibet (Section 5.1) often causes them to be convergent east of 20E. With moisture deeply spread through the troposphere and instability least, squall frequency decreases to a secondary minimum at Fort Lamy/Maiduguri in August. As this is also the wettest month, either the fewer squalls are rainier or a Zone D *rains* situation occasionally affects the station.

Zone D and *rains* prevail over the southern stations and squalls cease. Rainfall decreases to a secondary minimum at Lagos, but orographic lifting maintains high rates at Douala/Calabar at the foot of the Cameroon Plateau. The high-speed center in the upper-tropospheric easterlies is as likely to be a result of the rain (Raman and Ramanathan, 1964) as a cause (Koteswaram, 1958) (Section 3.4.2).

5.7.5   AUTUMN TRANSITION

Autumn resembles spring. The surface heat low and 200-mb ridge occupy the same latitudes, upper-tropospheric convergence is usually absent, and instability reaches a secondary peak. Squall frequency maxima are smaller than in spring. The sequence followed through the spring transition is not just reversed, but accelerated. By November, Fort Lamy/Maiduguri is once more in Zone A; the southern stations revert to winter conditions by December. The advance to midsummer takes about two months longer than the retreat to winter.

In October 1954 (Figs. 3.24 and 3.25), Fort Lamy/Maiduguri lay in Zone B. Middle-tropospheric subsidence overcame low-level convergence to maintain a rainless heat trough. To the south, greatest rainfall and low-level divergence nearly coincided—impressive evidence of the presence of subtropical cyclones (Section 3.5.6).

5.8   Asia

The Asian monsoons are complex and varied. Until recently, the Indian and Chinese branches were studied intensively but separately, and the Burma branch hardly at all. Weather satellites and aerological observations over southwestern China and Tibet have now made obvious the interdependence of all branches. I have already emphasized the dominant role of the Himalayan–Tibetan massif (Section 5.1).

The people of tropical Asia farm, but seldom in the arid winters. The drought-breaking rain of late spring or early summer is life giving and often life saving. Normal dates for the rain to start, shown in Fig. 5.17, are derived from authorities applying different criteria. Even if one ignores the large interannual variability, climatology scotches simplistic notions of steady progression of the rain. Over southern China, rainfall increases from mid-January; on the average over Indochina the rain begins in mid-May: Mandalay (21°59′N, 96°06′E) lags Akyab (20°08′N, 92°53′E), 400 km to the southwest, by three weeks.

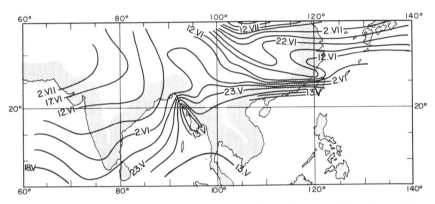

Fig. 5.17. Average onset dates of the rainy season (From Ramdas, 1949; Flohn, 1960a; Kao *et al.*, 1962; Huke, 1965).

Day-to-day changes are even more confusing, as Jawaharlal Nehru observed. Nevertheless, from the earliest days of monsoon research in Asia, meteorologists have constantly tried to find order in what is only superficially a very orderly phenomenon. They have written papers purporting to identify singularities in annual variation of rainfall and many more claiming that particular meteorological events can be used accurately to forecast onset or cessation of monsoon rain.

Order has not been summoned forth. Almost as many singularities have been specified for the spring rains of eastern Asia as there are days in spring, while the number of papers citing onset predictors is being overtaken by the number disproving the predictors or arguing about the definitions of " onset." As data have accumulated, so the systematists' quest has flagged despite the voracious appetite of meteorological consumers for more.

I shall not spend more time on these man-made complications. Each major, often sudden circulation and weather change accompanying the march of the seasons seldom recurs on about the same date nor are the same precursory signs regularly followed.

5.8.1   AUTUMN TRANSITION (OCTOBER, NOVEMBER)

Winter comes suddenly to the Himalayas. In autumn the polar westerlies begin to intensify and move south, reaching the western Himalayas in the second half of September. An eastward-moving trough in the westerlies may then interact with an end-of-season monsoon depression. The depression intensifies and recurves east of the trough axis, giving heavy rain over Punjab and Kashmir (Section 3.4.3) (Pisharoty and Desai, 1956).

The westerlies continue southward. Along the southwest face of the western Himalayas, lying athwart the flow, strong lifting develops in late September. The normal temperature gradient, locally strengthened by adiabatic cooling from lifted air over the mountains, presumably leads to establishment of a subtropical jet stream, although the precise dynamic mechanism involved is not clear (Ramage, 1952c). The jet, once established in the northwest (Reiter, 1959), rapidly produces conditions along the mountain barrier to the southeast which extend it in that direction at a rate of about 3° long, per day (Yeh, 1950). By mid-October the jet is established along the Yangtze Valley and to the south of Japan (Yeh et al., 1959). From then until the end of winter the thermally and mechanically anchored jet (Bolin, 1950; Staff Members, 1957–1958; Reiter, 1963; Miyakoda et al., 1969) seldom shifts from the Himalayas and the Yangtze. Most rain falls along the line of the jet (Yeh, 1950; Chaudhury, 1950) but is inhibited by subsidence to the south (Ramage, 1952c). Because of Himalaya–Tibet (Section 5.1.1), upper-tropospheric divergence and unsettled weather associated with the jet are persistent and extensive over China, but restricted and intermittent over the Himalayan foothills.

Appearance of the subtropical jet ends the Chinese "Indian Summer" (Kao, 1958; Section 5.8.7.6), which usually begins in mid-August. Western China and the Yangtze Delta experience more rain and rain days in October than in later months (Kao and Kuo, 1958). In October, typhoons and other tropical disturbances (Section 5.8.1.2) bring tropical maritime air beneath the subtropical jet, either from the China Seas or from the Bay of Bengal. Thus more precipitable water is available than later in the season when air masses over China are almost exclusively of continental origin.

Over Siberia the polar anticyclone is intensifying. By mid-October cool surges of the northeast monsoon have reached northern Indochina and southern China, and by the end of the month are cooling the surface of the central South China Sea (Fig. 5.7).

Surges do not penetrate south of the Himalayas. Consequently, insolation resulting from clear skies stalls the normal temperature fall, maintains moist inflow and a *showers* regime over the Indian and Indochina peninsulas and together with small evaporative cooling from light winds produces a secondary sea-surface temperature maximum in the Arabian Sea and the Bay of Bengal (Fig. 5.7).

North of 25N, winter has arrived, accelerated by the Himalayas. South of 25N, progress more sedately follows the sun. As the heat trough disappears over northern India and Pakistan, the near-equatorial trough redevelops close to 10N (Fig. 5.18). Above it deep easterlies prevail south of 20N.

Other facets of October weather are discussed in Section 5.8.4 where I compare it with May weather, the corresponding early-summer transition month.

In November weak anticyclones hold sway over northern India and Pakistan. The northeast monsoon prevails, feeble south of Tibet and often strong east of Tibet. Coming off the continent, where *showers* have practically ceased, the air is dry and cool. It picks up heat and moisture on its way to the near-equatorial trough along about 8N.

The mean vertical circulation in the transition blends winter and summer. A Hadley cell (Section 5.8.2) connects the zone south of the jet stream and the near-equatorial convergence while a monsoon cell or a Southern Hemisphere Hadley cell (Section 5.8.7.1) links the convergence with the Southern Hemisphere. Development is favored of weak surface disturbances, which may give

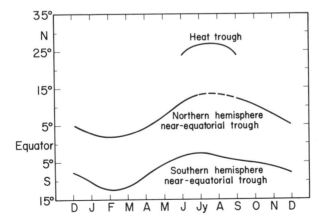

Fig. 5.18. Annual latitudinal variation of lower tropospheric (1.5 km) pressure troughs over the Indian Ocean (Raman, 1970). Meinardus (1893), by averaging ship observations, located surface troughs in almost the same latitudes.

torrential rain (Section 4.2.2.2). Together with westward-moving tropical storms (Section 5.8.1.2) these account for the India and Indochina east coast rainfall maxima, between 15 and 20N in October and south of 15N in November. The winter vertical circulation becomes dominant over the South China Sea during the last ten days of November (Section 5.8.2.4).

Weather is controlled by jet stream activity in the form of short-wave troughs or disturbances, by the near-equatorial convergence or by near-equatorial trough disturbances which sometimes become typhoons. In between, the region experiences fine weather, only rarely interrupted by northward-veering typhoons.

5.8.1.1   *Disturbances in the Polar Westerlies and Monsoon Surges.* Surface cyclogenesis is favored east of long-wave upper-tropospheric troughs near 20E and 130E (Fig. 5.2).

In the west, a weak surface depression ("western disturbance") develops over Iran where a cold surge from southern Russia meets warmer, moister air east of an upper-tropospheric trough (Datta and Gupta, 1967; Pisharoty and Desai, 1956). Western disturbances travel at about 10 m sec^{-1} on an average, in the direction of the 300 mb flow; some skirt the southern edge of the Himalayas. They last from 3 to 4 days. Accompanying rain is usually confined to the foothills. West of a disturbance center, cool dry air from north-western India may spread into central India behind a weak cold front (Section 5.8.2.2).

East of Tibet, short-wave troughs worsen weather as they pass over central China. However, they seldom break down the underlying anticyclone until they reach the east coast (Wang, 1963). There, the long-wave trough, jet-stream divergence, and large topographically anchored temperature and moisture gradients initiate surface cyclogenesis. West of the cyclone, and involved in the same development, the Siberian high intensifies (Section 3.1.3.1) and a cold front sweeps south on the leading edge of a monsoon surge.

The northeast monsoon is most persistent in November. The disturbances which trigger monsoon surges are generally weak and far to the north or east. Within the monsoon region, northeasterlies weaken and back slightly ahead of a cold front, then freshen and briefly veer as it passes. Transition and early winter disturbances help maintain the monsoons without interrupting the surface circulation.

5.8.1.2   *Tropical Disturbances.* Thirty-six percent of all Bay of Bengal and Arabian Sea tropical cyclones form in October (Fig. 5.35) and November (Ananthakrishnan, 1964). A vigorous surface trough and slight vertical wind shear favor development (Section 3.3.1).

Many of the cyclones, as well as less intense disturbances move across peninsular India. Of those that recurve northward across Burma, a few may initiate disturbances which cause heavy rain over central or southern China.

Typhoons and tropical disturbances cross the Philippines or develop in the South China Sea. Most move westward south of the upper-tropospheric ridge and occasionally cause weather to deteriorate as far away as southern China, more than 700 km north of the track (Cuming, 1968).

The monsoon determines the histories of typhoons crossing the China Seas north of 15N. During a lull in the monsoon, a typhoon may penetrate southern China or the Gulf of Tonkin. Two of the most severe and prolonged Hong Kong typhoons struck in October. A monsoon surge may intercept a typhoon which rapidly fills when cold, dry air reaches the eye (Ramage, 1956). Then again, some typhoons recurve across the subtropical ridge before being affected by surface cold air. Then, if the thermal gradient is large, a typhoon weakens only slowly as it races into higher latitudes. The reason is threefold: (a) the center moves too fast to be overtaken by surface cold air and so remains relatively warm-cored, and (b) too fast to induce much cold water to upwell to the surface; (c) the great vertical wind shear enhances convection, just as in squall lines (Section 3.11.1). Typhoons strengthen the northeast monsoon by increasing the pressure gradient. Usually, westward passage of a center to the south causes the wind to veer temporarily to east.

Between 1947 and 1962, 12 tropical disturbances moved west or northwest across the Gulf of Thailand during October and November (Montriwate, 1963). The six October disturbances were more severe than the six November disturbances. Two of them, which almost reached typhoon intensity, caused extensive wind and flood damage over southern Thailand before crossing the Kra Isthmus into the Andaman Sea.

5.8.2 WINTER (DECEMBER, JANUARY, FEBRUARY)

North of 25N, the circulation established in October intensifies. South of 25N the remnants of summer disappear. As the sun moves far into the Southern Hemisphere, the near-equatorial trough weakens and the northeast monsoon extends across the equator to become the northwest monsoon of Indonesia and Australia (Sections 5.5.1.4 and 5.9.2) and the northerly monsoon of eastern Africa and the Azanian Sea (Sections 5.4.2.3 and 5.9.2).

The Northern Hemisphere Hadley cell now overlaps 1000 km into the Southern Hemisphere. The southerly and southwesterly winds comprising the upper-tropospheric branch of the Hadley cell converge with westerlies to the south of the subtropical jet to produce persistent subsidence south of 25N (Ramage, 1952c). Weather remains unsettled along the jet-stream axis, especially across central China.

5.8.2.1   *Szechwan Cloudiness.* The Red Basin of Szechwan, about 16,000 km^2 in area, lies east of the southeastern corner of Himalaya–Tibet. Cold air which reaches the basin from the northeast tends to stagnate there during the late autumn and winter.

Air between 2000 and 5000 m just south of the Himalayas is constrained to flow from west, parallel to the range. On reaching the southeast corner of Himalaya–Tibet, it swirls cyclonically across the Yunnan Plateau and slides up the dome of cold air trapped in the Red Basin. Persistent cloud results, often thick enough to produce drizzle or light rain. Wintertime sunshine in the basin amounts to less than 20% of the possible (Chu, 1962).

Most of the time, the Red Basin cloud mass is indistinguishable from the general cloudiness associated with the diverging jet stream over central and eastern China. However, on 16 January 1967 it was dramatically evident (Fig. 5.21). All China was in the grip of a deep cold outbreak, which, because of intense subsidence, was clear of cloud (Sections 3.1.3.1 and 5.8.2.3). Cold dry air covered the Red basin below 750 mb (Fig. 5.19). Between 750 and 600 mb, warm moist air from south of the Himalayas glided up over the cold dome and became saturated. Above 600 mb, air flowed across Tibet from the west and was dried through mechanical subsidence off the edge of the Tibetan Plateau.

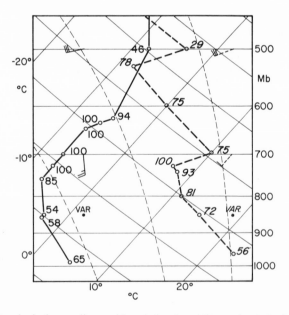

Fig. 5.19. Aerological soundings with relative humidity and winds (one full barb denotes 5 m sec^{-1}) for Ipin (28°49′N, 104°32′E) at 1200 GMT, 16 January 1967 (full lines) (Section 5.8.2.1) and at 1200 GMT, 28 March 1967 (dashed lines) (Section 5.8.3.2).

Where the northeast monsoon is vigorously lifted by the windward ranges in Taiwan, the Philippines, and Vietnam, skies are cloudy and appreciable rain falls. However, orographic lifting of this relatively stable air mass induces far less rain than near-equatorial trough disturbances in the autumn transition (Depperman, 1941) (cf. Sections 5.5.2 and 5.8.7.5). Elsewhere, south of the jet only a rare juxtaposition of disturbances or a major circulation change brings rain.

In the remainder of this section I first deal with the normal sequence of monsoon surges, then with superposition of tropical and extratropical disturbances, and finally with trough development in the subtropical westerlies.

5.8.2.2 *Monsoon Surges.* These become increasingly vigorous through January. Over India, a cold wave spreads southeastward as pressure rises behind a western disturbance (Mooley, 1957). The cold waves rarely reach south of 20N or east of 80E (Raghavan, 1967; Bedi and Parthasarathy, 1967) (Table 5.7). In one of the severest on record, minimum temperature over the Indian plains dropped 12C below normal.

Table 5.7.

Average frequencies (10-year period) of[a] (1) cold waves in central India[b] and (2) winter monsoon surges at Hong Kong[c]

		Sept.	Oct.	Nov.	Dec.	Jan.	Feb.	Mar.	Apr.	May
(1)	India									
	Gujarat			0.4	0.2	1.2	2.2	1.4	0.6	
	Central Maharashtra			0.2	0.6	1.4	2.6	0.6	0.4	
	Orissa						0.6	0.4		
(2)	Hong Kong	12	29	33	35	34	33	33	29	17

[a] The Indian and Hong Kong criteria are not directly comparable.

[b] After Raghavan (1967), the criterion being a drop $\geqslant 8C$ in minimum temperature below the normal minimum temperature.

[c] After Heywood (1953), the criterion being a shift to NE or N flow persisting or at least two days.

East of the Himalayas, surges are much stronger and more frequent (Table 5.7). Gales are common, while freezing temperatures have been recorded more than once at Hong Kong.

Over both India and China cold fronts leading the surges cause bad weather when they move beneath the jet stream. The latter, being topographically anchored, seldom shifts south with the fronts, which quickly dry out on reaching the subsident region south of the jet (Thompson, 1951; Ramage, 1952c).

5.8.2.3   *Nonfrontal Surges.* Generally, cold fronts can easily be followed as far south as 25N. But over southeastern China (Chin, 1969), the South China Sea, India (Ramaswamy, 1966), and northern Africa (Section 5.3.2.1), continuity can rarely be maintained.

Koo *et al.* (1958) found considerable variation in the supposedly conservative pseudopotential wet-bulb temperature at the upper boundaries of cold fronts over China. They concluded that the frontal surfaces could not be simple material surfaces, even on the synoptic scale, but that they cycled continuously through frontogenesis and frontolysis. Of 13 well-marked cold fronts which crossed the southern China coast during December 1957 and

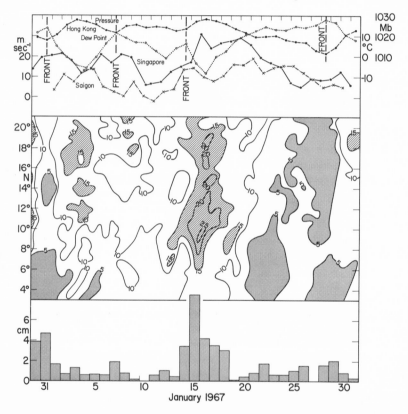

Fig. 5.20. Daily values for January 1967 over the South China Sea. Sea-level pressures (dashed line) and dew points (dotted line) at Hong Kong (22°18′N, 114°10′E), southerly components of the 200-mb winds at Saigon (10°49′N, 106°40′E) (dot–dashed line), 200-mb wind speeds at Singapore (1°21′N, 103°54′E) (full line), isopleths of surface wind speed (m sec^{-1}) averaged zonally between 110 and 118E, and histogram of rainfall averaged for five coastal stations [Kuching (1°29′N, 110°20′E), Bintulu (3°12′N, 113°02′E), Miri (4°23′N, 113°59′E), Labuan (5°21′N, 115°13′E), and Jesselton (5°57′N, 116°03′E)] in North Borneo.

January 1958, 10 originated or regenerated south of 30N (Chin, 1969). Raghavan (1967) reported similar developments over India.

When *vigorous* cold fronts reach 20N, northerly winds sometimes freshen almost simultaneously over several hundred kilometers to the south, well beyond where the winds behind it could have pushed the front.

Example: South China Sea, January 1967. Features of the monsoon over the South China Sea are shown in Fig. 5.20.

At Hong Kong, pressure rises and dew-point falls signaled cold front passages on 31 December, and 7, 14, and 28 January. As rough indicators of the strength of the Hadley cell, I have used the 200-mb wind speeds at Singapore and the southerly component of the 200-mb winds at Saigon (upper branch), the zonally averaged surface-wind speeds (lower branch), and North Borneo coastal rainfall (upward branch).

After the frontal passage on the 7th, winds increased as far south as 16N. An increase south of 8N and a slight rainfall peak may have been connected with the surge.

The mid-latitude situation preceding the frontal passage on the 14th has been discussed in Section 3.1.3.1. Dramatic changes took place over the South China Sea as well. In the north and the south, winds began to freshen on the 12th, two days before freshening began between 13 and 15N. At 200 mb, major increases occurred on the 12th at Singapore and on the 13th at Saigon. Rain was much heavier at North Borneo stations on the 14th than on the 13th.

By the 16th all indicators confirmed that the Hadley cell was intense (Fig. 5.21). Except for Szechwan (Section 5.8.2.1), China was clear of cloud, dominated by very cold subsiding air (Danielsen and Ho, 1969) (Section 3.1.3.1, Fig. 3.2). Offshore, the ocean rapidly added heat and moisture. On an 18-hr trajectory between Hong Kong (22°18'N, 114°10'E) and the Paracels (16°51'N, 112°20'E), the mixing ratio of the surface air increased from 1.2 to 8.6 gm kg^{-1}. Convective clouds grew to and spread beneath a sharp inversion at 2000 m and gave light showers. Farther from the continent, where the heat- and moisture-addition rates had moderated, the cellular, convectional nature of the clouds became apparent. Over and leeward of the high ranges of Taiwan, Hainan, and the Philippines skies remained clear. East of the Philippines, the cloud edge coincided with the cold air edge between 10N, 125E and 20N, 138E. Heavy clouds covered eastern Malaya and North Borneo, although between 3 and 8N, clouds were thinner and winds weaker than farther north.

Contemporaneously with the 31 December and 28 January frontal passages at Hong Kong, North Borneo rainfall and winds north of 8N increased.

Obviously, frontal movement and advective processes cannot directly

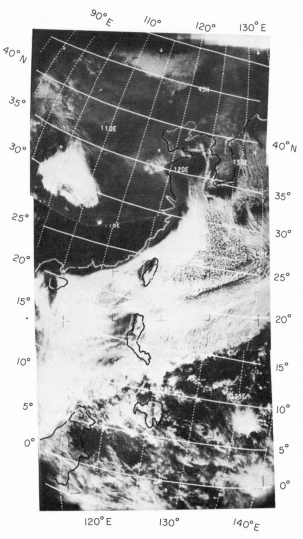

Fig. 5.21. Clouds photographed from ESSA 3 between 0530 and 0730 GMT, 16 January 1967 (Orbit Nos. 1328 and 1329).

account for these and other (Section 3.9.3) near-simultaneous intensifications of the Hadley cell, or of the vertical circulation of the summer monsoon (Section 5.8.7.1). I rather favor the hypothesis that the mechanism which initiates a surge also deepens the heat sink in the north. The increased south–north temperature gradient then accelerates the Hadley cell. *This* situation is what enhances North Borneo rainfall, although increased orographic lifting may play a contributory role.

Murakami (1969) devised a simple linearized system for eliminating gravity waves in a two-level, model troposphere. Beginning with the vorticity field determined from observations, he proceeded through the balance equation to an estimate of the initial pressure field, whence the initial divergence and vertical motion could be determined. When the primitive equations of motion were then applied in 10-min time steps to the initial pressure field (Ho and Murakami, 1970), the computation remained stable after two days, and physically reasonable developments resulted, even at the equator. According to the model, energy injected into the troposphere poleward of 20° results in an energy flux which can be detected at the equator within *12 hr*.

5.8.2.4 *Onset of Winter*. Bell (1969) remarked that in each of the five years from 1962 through 1966, the first simultaneous freshening of northerlies throughout the South China Sea occurred between 18 and 29 November. Thus would its lower-tropospheric branch respond to a Hadley cell expanding and intensifying to winter size and strength.

Data, particularly 5-day means, are insufficient to show whether the change always takes place in the last 10 days of November. Indirect evidence suggests that it might—the 50-yr frequency of typhoons over the South China Sea between 1 and 10 November differs little from the frequency between 11 and 20 November, whereas there is a drastic diminution in the 21 to 30 November interval.

5.8.2.5 *Superposition of Tropical and Extratropical Disturbances*. During winter the Northern Hemisphere near-equatorial trough sometimes becomes active and weak disturbances pass westward along it. They give some rain south of 12N. Rain from eastward-moving disturbances associated with the jet stream is confined north of 23 to 25N.

Rarely, relatively vigorous tropical and extratropical disturbances approach the same longitude and mutually intensify (Riehl and Shafer, 1944). Upward motion east of the northern disturbance is superposed on upward motion east of the southern disturbance, destroying the normal subsidence between 12 and 23N (Ramage, 1955; Pisharoty and Desai, 1956). The resulting bad weather is dissipated when the disturbances draw apart and subsidence returns.

Example: India, December 1962. (See Ranganathan and Soundarajan, 1965.) After moving across the southern peninsula on the 1st, a weak tropical depression became almost stationary off the west coast, just east of the longitude of a well-marked trough in the subtropical westerlies, also almost stationary. From the 2nd through the 4th, both systems intensified. Rising motion in lower-tropospheric southeasterlies and upper-tropospheric south-westerlies deepened the moist layer over the pensinsula and widespread rain ensued (Figs. 5.22 and 5.23).

At Bombay (Fig. 5.24), winds below 850 mb veered from northeast to southwest on the 3rd. Aloft, southwesterlies freshened in the northern disturbance west of the station. Early on the 5th the systems attained peak intensity, with surface southeasterlies overlain by strong southwesterlies, and moisture spread deeply through the troposphere. Thereafter, winds weakened and the moist layer contracted. The trough in the subtropical westerlies crossed the station late on the 5th as the disturbances separated; the northern moved east and the southern moved west and both rapidly weakened.

In Bombay, 48 mm fell in just over 24 hr on the 4th and 5th. The Bombay December average is 2 mm, and 4 out of 5 Decembers there are completely dry.

5.8.2.6 *Troughs in the Subtropical Westerlies.* Disturbances in the polar westerlies have already been discussed. Troughs sometimes originate south-west of India in the upper-tropospheric branch of the Hadley cell (Ramage,

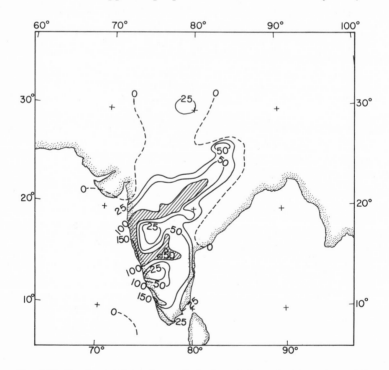

Fig. 5.22. Total rainfall (mm) over India from 1 through 6 December, 1962.

1955) or are identified crossing northern Africa. As the troughs move eastward across India, the weather remains fair, although winds above 400 mb veer from southwest to northwest. Between 90 and 100E the troughs usually intensify, possibly from superposing on the orographic trough caused by the southward bulge of the Himalayas, possibly because the long-wave trough pattern south of the mountains favors intensification in these longitudes. The troughs frequently become stationary over Thailand or Indochina or at other times troughs apparently develop in these longitudes. They behave like any quasi-stationary trough in the polar westerlies, with upward motion to the east and subsidence to the west of the trough line.

Any front lying across the northern part of the South China Sea is activated. In the lower troposphere, winds veer rapidly with height and glide up the frontal surface. If the trough is weak, no more than a layer of stratus with drizzle or light rain results. If the trough is vigorous, disturbances with considerable rain areas extending to the north move east–northeast along the front and often develop into surface depressions over the East China Sea. North of southern China, normal conditions prevail, with the disturbances to the south helping to maintain a rather stronger than normal monsoon.

Troughs in the subtropical westerlies generally dissipate *in situ* when the upper-tropospheric branch of the Hadley cell is restored and reestablishes normal subsidence south of the jet stream.

This situation has rarely been observed before mid-January. It then becomes increasingly frequent and in spring accounts for much of the enhanced cloudiness and rain over southern China and North Vietnam (Section 5.8.3.2). McCutchan and Helfman (1969) described a spring example.

The reason for development and dissipation of these troughs escapes me. Temporary breakdowns of the Hadley cell presumably are related to large-scale equator-spanning changes—circulation adjustments to the waning of winter.

Example: February 1968. On 2 February a cold front moved slowly southward across the southern China coast and then remained stationary until moving away southeastward on the 8th. The situation on the 6th typifies conditions which remained relatively unchanged for more than a week (Figs. 5.25, 5.26, and 5.27).

At the surface the front lay across the northern part of the South China Sea, but was difficult to locate off Vietnam, where northerlies, channeled along the coast, had spread beyond 15N (Navy Weather Research Facility Staff, 1969). The northeast monsoon circulated around a ridge extending southeastward from an anticyclone over Mongolia.

At 300 mb (Fig. 5.25), the trough in the subtropical westerlies lay along

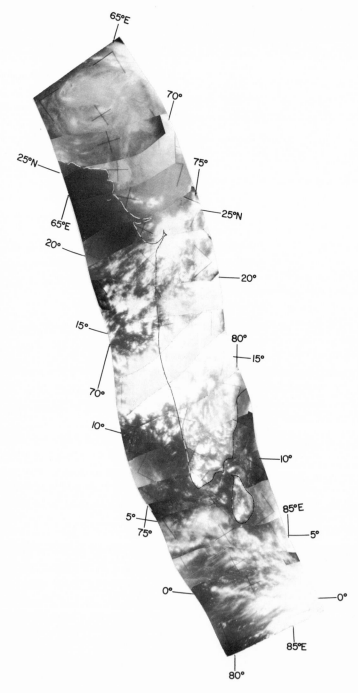

Fig. 5.23. Clouds photographed from TIROS VI at 0900 GMT, 4 December 1962 (Orbit No. 1123).

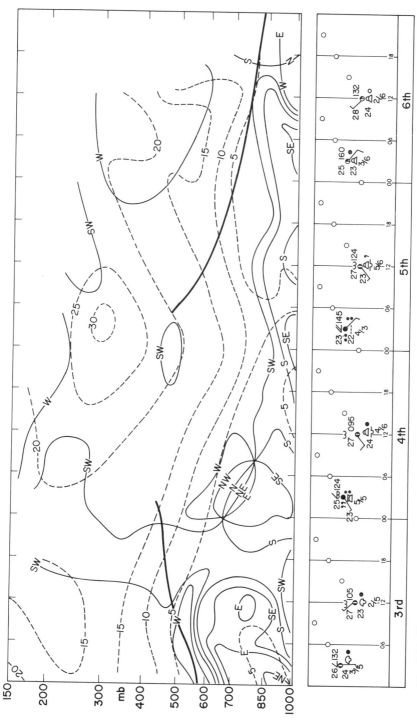

Fig. 5.24. Time cross section for Bombay (19°07'N, 72°51'E) from 3 to 6 December, 1962. Isogons are shown full, isotachs (m sec^{-1}) dashed, and the upper limit of relative humidities above 50% thick and full.

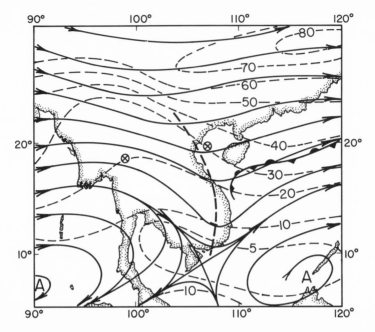

Fig. 5.25. Circulation at 300 mb for 0000 GMT, 6 February 1968. Isotachs are labeled in meters per second. Locations of the surface front and of the soundings shown in Fig. 5.27 are indicated.

Vietnam. *West of the trough* convergence caused strong subsidence to below the lifting condensation level of surface air (Fig. 5.27).

East of the trough divergence merged with the usual jet-stream divergence east of Tibet to produce ascent and an extensive cloud sheet (Fig. 5.26). In the south, upglide along the front and heat and moisture, added to the cold air by the coastal waters, almost saturated the air below and above the front (Fig. 5.27), and crachin (Section 5.8.3.2.) occurred intermittently. Nevertheless, rainfall was slight.

Above 600 mb, dry air passed through the trough with only slight cooling. Between 800 and 700 mb, mere lifting of the air previously west of the trough could not have produced the high humidities observed to the east.

Moisture must have been added from the South China Sea. Thus, compared to a weak trough, a vigorous trough would increase cloud depth to the east both by causing greater upward motion and by increasing the southerly component of the wind, thereby tapping additional moisture.

The sharp western edge of the cloud, a winter climate feature, is of orographic origin, marking the limits of lifting condensation over the Szechwan cold pool and on the windward slopes of the Annam Mountains (Section

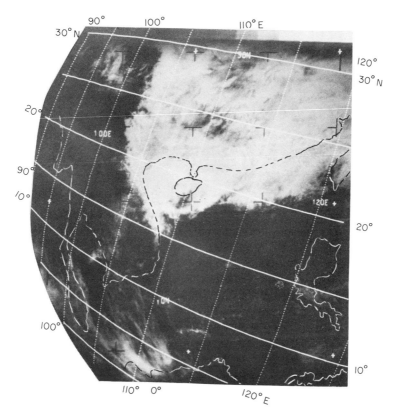

Fig. 5.26. Clouds photographed from ESSA 3 at 0500 GMT, 6 February 1968 (Orbit No. 6176).

5.8.2.1). Upper-tropospheric convergence to the west and upper-tropospheric divergence to the east of troughs in the subtropical westerlies may, as in this example, accentuate the effect. There was apparently no snow over southeastern Tibet (Fig. 5.26) (cf. Flohn, 1968).

Between the 7th and 8th the trough in the subtropical westerlies moved eastward and became very weak. The cloud sheet thinned and broke.

Long-wave trough retrogression and sea fog along the South China coast and in the Gulf of Tonkin may occur at any time after mid-January, but since they are more typical of spring than of winter they will be discussed in the following section.

5.8.3   Spring Transition (March, April)

The shift from winter to summer monsoon starts in March in the surface layers and extends throughout the troposphere by late June (Kao *et al.*, 1962).

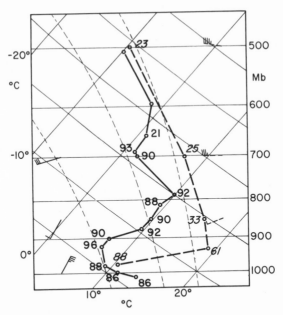

Fig. 5.27. Aerological soundings with relative humidities and winds (one full barb denotes 5 m sec⁻¹), at 0000 GMT, 6 February 1968. (1) West of trough in subtropical westerlies, Chiangmai (18°47′N, 98°59′E) (dashed lines); (2) east of trough in subtropical westerlies, U.S.S. Belknap (19°30′N, 107°00′E) (full lines) (Navy Weather Research Facility Staff, 1969).

Aloft, the winter circulation weakens but does not materially change during March and April, with the subtropical ridge displaced farthest south in March.

The near-equatorial trough lies below 5N and is weak. Tropical storms are almost unknown in the Arabian Sea and are very rare in the Bay of Bengal and the South China Sea. Those which do develop generally recurve near 15N.

5.8.3.1 *West of 100E.* In March, surface northeasterlies still prevail over the Arabian Sea and the Bay of Bengal. Nevertheless, the winter monsoon has ended south of the Himalayas. Over India and Pakistan, south of 25N, average temperatures exceed 25C, enough to shift anticyclonic cells from land to the northern parts of the Arabian Sea and Bay of Bengal. The cells become more pronounced in April. By this time the subcontinent is a seat of relatively low pressure and heat lows are centered over peninsular India and Burma. Although vigorous land and sea breezes disguise the circulation, moisture is brought inland. Remnant middle-tropospheric subsidence keeps the weather hot, humid, and sunny. The lapse rate is conditionally unstable and afternoon thunderstorms may develop—a *showers* situation (Section

4.1.2). Typically, on 2 April 1967 (Fig. 5.31), scattered showers occurred over the Peninsula between 15 and 20N.

Despite the fact that the Arabian Sea and the Bay of Bengal are cooler than the land, sea-surface temperatures are highest in April, when clear skies favor insolational heating and light winds reduce evaporative cooling (Fig. 5.7). The seasonal increase in land–sea temperature gradient is thus retarded and this in turn hinders development of the summer monsoon circulation (Section 5.8.4.2).

In the north, near the jet stream, two regimes hold sway.

(*i*) *West of 85E.* Rainfall from western disturbances is greatest in March and diminishes in April and May as the jet and the disturbances weaken.

(*ii*) *Example.* On 2 April 1967 a weakening western disturbance was crossing Kashmir (Fig. 5.31) where it had caused thunderstorms. The associated cold front extended southwestward to the east of the Great Indian Desert, giving only scattered showers.

(*iii*) *East of 85E.* Rainfall increases steadily through March, April, and May. In these longitudes the Himalayas and the jet stream are closest to the sea. Because of this, strengthening winds from off the sea increase the supply of moisture for energizing western disturbances and enhancing rainfall (Flohn, 1968) (Section 5.8.4).

Over the region as a whole, April is the least cloudy month, a consequence of subsidence, weak surface circulations, and thermal and moisture equilibrium between sea and air.

5.8.3.2   *East of 100E.* Heat lows may develop briefly over southwestern China, but continued cool incursions east of Tibet help prolong the winter monsoon through March and most of April (Hong Kong is more than 10C cooler than Calcutta). The Siberian anticyclone is no longer monolithic; surface depressions cross China, where the jet stream still accelerates downstream. High-pressure cells follow the depressions eastward. In response, winds south of 25N fluctuate between northeast and southeast, with the latter predominating by the end of April (Heywood, 1953).

The great Northern Hemisphere Hadley cell gives way to a Hadley cell and a near-equatorial convergence in each hemisphere. Surges rarely invade the South China Sea where increasing insolation and diminishing winds replace cooling by heating at the sea surface (Fig. 5.7).

Between the near-equatorial convergence and South China coastal waters, April is the finest month. Along the windward mountains of Taiwan, the Philippines, and Vietnam, where diminishing orographic lifting counteracts increasing moisture supply, rainfall tends to decrease; inland Vietnam, Cambodia, and Thailand experience a similar *showers* regime to peninsular India (Section 5.8.3.1) and increasing rainfall.

Weather is changeable over southern China and its coastal waters. Fronts, which often stagnate here in the polar front zone, are activated by troughs in the subtropical westerlies (Section 5.8.2.6). As the year advances, increasingly moist air from over the steadily warming South China Sea glides up the fronts. Rainfall correspondingly increases and thunderstorms may occur. North of the surface front this situation maintains a fresh monsoon. In two other related situations, crachin and the West China trough, the monsoon is often interrupted.

(*i*) *Crachin.* Bruzon and Carton (1930) defined the crachin of the coastal regions of northern Indochina and southern China as

"... a humid period of fogs and drizzle or light rain which sets in at about the time of the normal annual temperature minimum, generally toward the end of January, and interrupts the dry season ... Then, even though temperatures rise, the crachin may persist into mid-April, gradually merging with the rains of the rainy season proper."

Crachin, the most important coastal bad-weather regime of the cool season, is a low-level stratus phenomenon, most clearly marked in the arid belt between the jet-stream center (28N) and the subtropical ridge (17N) where indeed it predominates in late winter and spring. Though there may be prolonged precipitation, the amounts recorded are small. Visibility may be rapidly and seriously reduced at its onset.

Crachin may develop in two ways:

(1) As the result of mixing of two nearly saturated air masses along a frontal surface. Mixing occurs only during or after passage of a cold front, and, as Petterssen (1939) pointed out, raises the cloud base to near the top of the cold air which is usually far from saturated.

When a trough in the subtropical westerlies lies over Vietnam, warm air sliding up a front off southern China may cause rain (Section 5.8.2.6). Evaporation from the rain and moisture added from the sea then combine to form low-based cloud and to reduce visibility within the cold air (Fig. 5.26). Sea fog is prevented by the sea being warmer than the air.

(2) By surface cooling of a warm moist air mass. This is almost always the cause of the worst and most persistent crachin (Ramage, 1954).

Either of two readily identifiable synoptic sequences may be associated with this nonfrontal crachin. One, which is most important in late winter, begins when a continental anticyclone previously consisting of polar continental air moves eastward from the mainland. Air on its eastern side follows a track across the warm waters south of Japan before swinging west to cross the cold waters along the China coast (Fig. 2.2). There, rapid cooling and turbulent mixing often result in crachin development. First, a layer of stratus forms beneath the turbulence inversion. Then, as the moisture content of the incoming air steadily increases, the stratus builds downward, drizzle sets in and eventually sea fog may form. Should the anticyclone center become almost stationary to the south of Japan, intense and persistent crachin develops which can be dispersed only by a fresh surge of the monsoon.

The other sequence which becomes increasingly important as spring advances usually begins when a trough in the westerlies intensifies over western China (see below). Surface pressure falls over the mainland and a wedge of the Pacific anticyclone extends across the Philippines to China. This brings tropical maritime air to the region. Temperature contrasts between the advected air and the coastal zone are sharp and crachin often takes the form of dense sea fog. Over land, marked diurnal variation of crachin may occur, for when the anticyclone center lies far to the east and pressure gradients are small, mechanical turbulence is insufficient to counteract the effect of increasing temperature during the day and the stratus cover may be temporarily dispersed.

Nonfrontal crachin develops only in air which has first been strongly heated over warm seas and then cooled near the coast. As would be expected, it is confined to the coastal belt of cold water, being seldom observed more than 150 km offshore. Sverdrup (1945) and Burke (1945) have reported that air heated over the oceans usually attains a fairly stable surface relative humidity of about 80% after some 800 km of sea track, and a surface temperature but little different from the sea-surface temperature. Thus, through a normal crachin period when trajectories of air reaching the coast swing progressively farther southward over warmer and warmer water, a steadily rising surface dew point is observed along the coast. At Hong Kong the rise usually amounts to about 2C in 24 hr until the dew point approaches to within 1C of the sea temperature. This and the fact that the dew point varies little as the air crosses the coastal waters indicates that no significant condensation occurs at the surface until the fog point is reached. Since at the same time the dry-bulb temperature reacts rapidly, falling to within a degree or two of the sea temperature, the condensation level drops, reaching the level of turbulent mixing, and crachin forms.

The cold sea cools and *dries* the air, for not only heat but also moisture is transported down to the sea surface where it condenses (Roll, 1965). Fog may never form unless other factors, particularly radiative cooling, are operating (Emmons and Montgomery, 1947).

Even with sufficient surface and radiative cooling, fog will not develop if turbulent mixing brings dry air down to the surface (Section 5.8.7.4.) Along the China coast and in the Gulf of Tonkin the subsidence inversion is usually well above the top of the turbulence layer and radiational cooling and turbulent mixing suffice to produce sea fog once the dew-point and sea-surface temperatures coincide.

Crachin becomes increasingly frequent through winter and spring (Table 5.8).

Table 5.8
Drizzle and fog frequency through the cool season at Hong Kong

	Nov.	Dec.	Jan.	Feb.	Mar.	Apr.
Mean percentage of hours with drizzle in each month (26 years)	7	8	10	15	17	17
Mean percentage of days with fog in each month (60 years)	3	7	13	18	28	25

Frequencies of troughs in the subtropical westerlies (frontal crachin) and of West China troughs (nonfrontal crachin) increase correspondingly.

(*ii*) *West China trough.* Between October and April, 130E marks the average longitude of a major long-wave trough in the polar westerlies (Fig. 5.2). Surface cyclogenesis usually takes place east of this meridian. A few times every cool season and usually after mid-January, the trough along 130E dissipates or moves eastward. A warm wedge extends westward from the Pacific anticyclone bringing fine weather to South China and sea fog to the China coast. The Siberian anticyclone retreats westward, and over southern China dew point and temperature rise and pressure falls. At this stage, a trough in the westerlies moving east from India will intensify over western China. This results in the region of surface cyclogenesis being displaced about 20° westward. As long as the trough remains over western China, central China and sea areas from the Formosa Strait northward experience extremely unsettled weather.

Conditions in the Red Basin of Szechwan are transformed from the usual drizzling overcast, typical of the cool season (Section 5.8.2.1). Warm air from the south displaces the usual stagnating cold air. Aloft, air moving through the West China trough has recently subsided and is still dry when it reaches Szechwan and so, in the absence of the dome of cold air in the basin, clouds become scattered, and a heat low develops there. Heat lows and

West China troughs appear to be climatologically associated; from 1955 through 1958, 20 heat lows developed, 16 of them in February, March, or April. In those same months there was a heat low over the Red Basin on 15% of the days as compared with 2% of the days during the remaining months (Li, 1965).

When hemispheric flow returns to normal the West China trough moves eastward. The wedge from the Pacific anticyclone has earlier brought tropical maritime air far inland and warm-front and cold-front thunderstorms are common. West of the trough line there is vigorous subsidence, an anticyclone builds behind a surface cold front, which displaces the Szechwan heat low, and normal conditions return. Just ahead of and at the trough, the surface front and depression system may travel sufficiently far south to give rain to southern China and northern Indochina, although south of 20N weather stays fair.

Occasionally a West China trough situation lasts from 10 to 20 days. The trough and its associated depression may move eastward. However, there is no rapid anticyclogenesis west of the trough line. Instead, pressure falls again over China and upper winds back from northwest to southwest. The eastward-moving trough weakens as a new trough intensifies over western China. While it remains stationary it is a source of shallow disturbances traveling east-northeast. South of the disturbance track, southern China experiences persistent southerly winds and unseasonably warm and humid weather. When the trough finally moves eastward, the continental anticyclone is restored to its usual position.

(iii) Example: March to April 1967. On 25 March the first of a succession of eastward-moving short-wave troughs in the subtropical westerlies intensified over western China. For the next week, each trough, as it continued eastward, weakened rapidly.

A series of short-wave troughs in the polar westerlies independently moved southeastward across southern Siberia and Manchuria. These troughs caused cyclogenesis along the polar front which did not move south of 25N until the 29th. The troughs in the north moved faster than the troughs in the south and so were rarely in phase with them (Chow and Koo, 1958).

The Japan Meteorological Agency (1967) prepared 5-day mean charts for 27 through 31 March 1967. The 500-mb pressure-height anomalies were negative over western China and positive south of Japan, while the Siberian anticyclone was centered 2500 km westnorthwest of its normal position.

At the surface, a wedge from the Pacific anticyclone brought warm moist air to southern China and North Vietnam. Air in the middle troposphere approaching the West China trough from the west underwent dynamical subsidence, and, as it left eastern Tibet, mechanical subsidence as well.

Decreasing density of the air column and reducing cloudiness combined with *warm lower layers* to transform the normal Red Basin cyclonic swirl, in which cloud forms in warm air lifted over a *cold* dome (Section 5.8.2.1), into a heat low (Fig. 5.19) centered near Kweiyang (26°34′N, 106°42′E) (Ables, 1969). This accentuated inflow from over the sea; winds in the Gulf of Tonkin often exceeded 12 m sec^{-1}.

The situation at 0000 and 0600 GMT 28 March 1967 typifies the period (Figs. 5.28–5.30). Two distinct trough and jet-stream systems are evident, not uncommon in spring when the subtropical jet is weakening and the polar jet is intensifying. The West China trough was rather weak. Dry subsided air remained almost cloud free over the heat low and up to 800 km beyond the trough. At 0600 GMT patches of fog persisted along the South China coast and along and off the Red River. Clouds appeared denser north of the polar front than south of it.

Sea fog developed along the South China coast by the 26th and persisted until the 30th, undergoing considerable diurnal variation, even some distance offshore. A ship stationed in the Gulf of Tonkin made twice-daily rawin

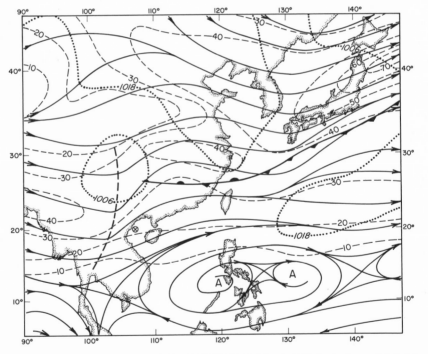

Fig. 5.28. Circulation at 300 mb for 0000 GMT, 28 March 1967. Isotachs are labeled in meters per second. A selection of surface isobars is shown as dotted lines. Locations of the surface front and of the soundings shown in Fig. 5.30 are indicated.

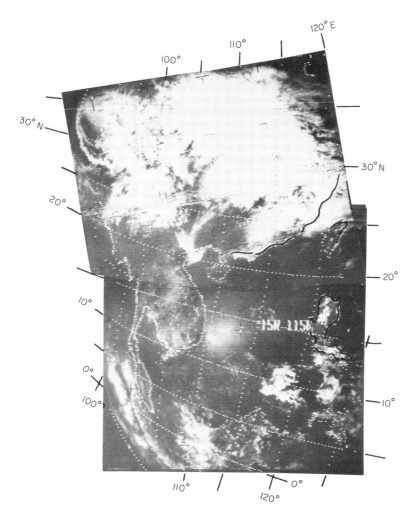

Fig. 5.29. Clouds photographed from ESSA 3 at 0600 GMT, 28 March 1967 (Orbit No. 2220).

soundings. Those for the 28th are typical (Fig. 5.30). At 0000 GMT (0700 hr, local time) base of a subsidence inversion was about 300 m above the surface and the weather was foggy. At 1200 GMT (1900 hr, local time) the inversion coincided with the surface and the fog had dissipated. Presumably as the heat low intensified during the day, inflow accelerated. Ensuing divergence over the gulf could have brought the subsidence inversion down to the surface where the dry air would have prevented fog formation. This pattern prevailed elsewhere along the coast where the fog largely dissipated

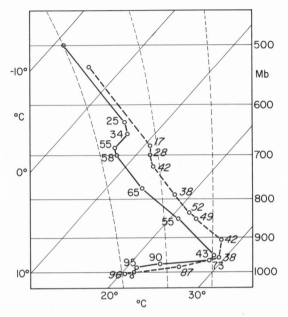

Fig. 5.30. Aerological soundings with relative humidities for 20N, 107E, at 0000 GMT (full line) and 1200 GMT (dashed line), 28 March 1967 (Navy Weather Research Facility Staff, 1969).

during the afternoon. Up to 50 km offshore, vertical circulations in a land breeze/sea breeze cycle would also lift the inversion at night and lower it during the day.

At the end of March a trough in the polar westerlies linked with the West China trough and both moved eastward as the long-wave pattern reverted to normal. In the process, the polar front shifted to southern China, where it caused heavy rain and some thundershowers on the 1st and 2nd of April (Fig. 5.31).

5.8.4   EARLY SUMMER TRANSITION (MAY)

Just as over northern Africa (Section 5.7), transition from the winter to the summer monsoon (against the normal meridional temperature gradient) takes longer than the reversion from the summer to the winter monsoon.

Over Asia, May should resemble October, but as Fig. 5.32 shows, there are significant differences. May is wetter over China and October is wetter over India, while, except for the Andaman Sea, October is wetter over the Bay of Bengal and the South China Sea. These differences can be related to circulation differences between the two months, suggesting in turn why summer monsoon rain starts earlier to the east than to the west of 90E.

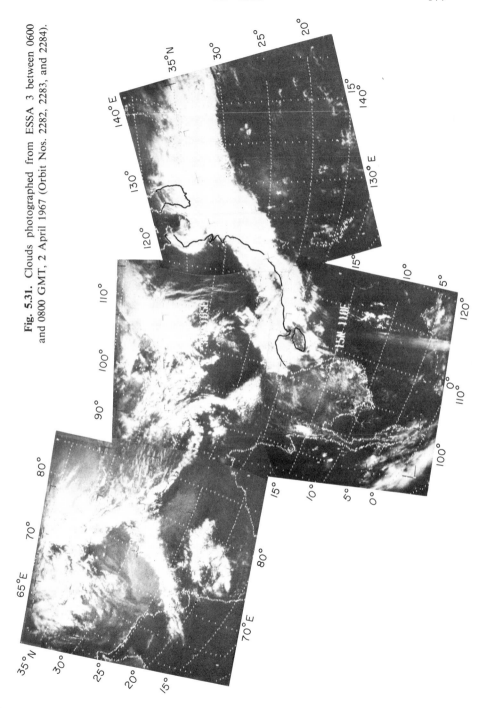

Fig. 5.31. Clouds photographed from ESSA 3 between 0600 and 0800 GMT, 2 April 1967 (Orbit Nos. 2282, 2283, and 2284).

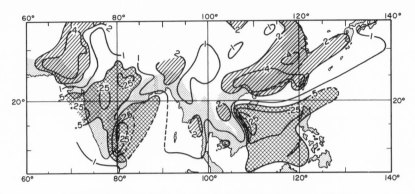

Fig. 5.32. Ratio of May mean rainfall to October mean rainfall.

5.8.4.1 *May and October over China.* When the winter monsoon sets in, coastal waters cool as well as the land. By deepening the surface mixed layer of the ocean, by evaporation, by conduction, and by generating a southward-setting current, the strong dry winds so enhance the radiational cooling effect that along the South China coast the fall in sea-temperature lags the fall in inland air temperature by less than a month. Conversely, during the spring transition, light, moist winds result in a shallow mixed-layer and slight evaporational cooling. Thus radiational heating raises surface-water temperatures at the same rate as air temperatures increase inland.

North of 40N, strikingly different regimes prevail. From October to January, the temperature difference between the sea surface and air over the land increases to a maximum. The Sea of Okhotsk begins to freeze in November, releasing latent heat of fusion which, by slowing the cooling rate, enlarges the sea–land temperature difference. During the spring transition, melting ice, by absorbing heat, keeps the water cold so that by May overland air temperatures are far higher than sea-surface temperatures. The tropospheric circulation responds. Average *surface* pressure gradients, directed toward the sea in October and through the winter, reverse by May (Kurashima, 1968) and the "Okhotsk high" develops (Okada, 1910), locally *accelerating* transition to a summer circulation.

The new offshore cold center begins to affect the *middle and upper troposphere* as the thermal–mechanical influence of Himalaya–Tibet on the subtropical westerlies is waning. The long-term monthly mean pressure-height profiles along 35 to 40N (Figs. 5.2 and 5.33) show the winter long-wave trough shifting east of 130E during April and May. It also weakens at the same time as the West China trough is becoming a significant average feature.

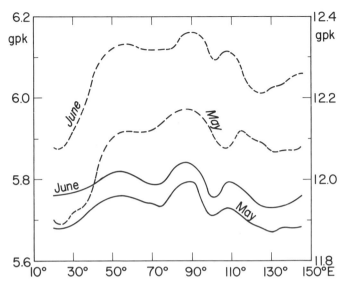

Fig. 5.33. Longitudinal variation of mean monthly pressure heights in geopotential kilometers at 500 mb (full lines) and 200 mb (dashed lines) in the zone 35–40N for May and June.

Below 850 mb, a wedge from the Pacific anticyclone now covers the northern part of the South China Sea and often extends across southern China. As coastal waters are warming, crachin seldom forms after the first week in May in the south or after the end of May along the East China coast. Tropical maritime air from the Pacific meets polar continental air from central China (Staff Members, 1957–1958) or Polar maritime air from the south side of the Okhotsk high (Okada, 1910) along the polar front which fluctuates north and south across southern China (Heywood, 1953; Yoshino, 1963) (Fig. 5.36). Maximum rainfall occurs north of the surface trough (Fig. 5.34), suggesting frontal slope. The West China trough and divergence in the subtropical jet over China often activate the polar front.

Warm air penetrates western China, although the Red Basin heat low can rarely be identified as a separate entity after April (Li, 1965). Chengtu (30°40′N, 104°04′E) is 3.8C warmer and pressure is 10.1 mb lower in May than in October.

By May the combined action of a heat low south of the Himalayas and the West China trough draws warm moist air from the Bay of Bengal into northeastern India (Section 5.8.4.2). Thence, it is brought across the Yunnan Plateau between 850 and 600 mb into southern China (Murakami, 1959b) (see Fig. 5.37). Thus, south of the polar front, moisture is supplied from both

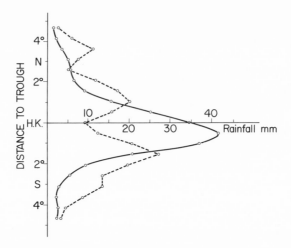

Fig. 5.34. Mean daily rainfall at Hong Kong (22°18′N, 114°10′E) associated with nearby surface troughs (1958–1967): May and June, full line; July and August, dashed line (Bell, 1969b).

the Pacific and Indian oceans, and, although disturbances forming on the polar front may have weak circulations, they produce extensive and sometimes torrential *rains* (Section 4.2.2.2). In some years, in the first half of May, the rain belt moves abruptly northward over the Yangtze Valley a month before the usual onset of the plum *rains* (Mei-Yü) (Hsu, 1965) (Section 5.8.6.2).

5.8.4.2   *May and October over Southern Asia.* On the average, locations of subtropical jet and near-equatorial trough and the *showers* regimes between them differ insignificantly in May and October. By the end of May a summer-like circulation (Section 5.8.5) develops. Intense heating has so reduced pressure over Arabia, West Pakistan, and northern central India that the anticyclones over the Arabian Sea and Bay of Bengal and the heat low over peninsular India give way to westerlies or southwesterlies.

After some delay in April (Section 5.8.3.1) the change to summer monsoon is locally accelerated over the western Arabian sea. As in the Sea of Okhotsk (Section 5.8.4.1), relatively low sea-surface temperatures are the cause. Once southwesterlies set in south of the heat trough, they increase evaporation and deepen the surface mixed layer of the ocean. However upwelling off Arabia and Somalia contributes most to surface cooling (Verploegh, 1960). This increases the temperature gradient and consequently the pressure gradient between land and sea (Section 2.2.4).

Over West Pakistan, although middle- and upper-tropospheric circulations and incidence of western disturbances are about the same in October, as in May, May is much wetter (Fig. 5.32). Because the heat trough is intensifying in May and disappearing in October, fresh southwesterlies over the West Arabian Sea in the former month bring significantly more moisture into the interior with consequence to the rainfall.

In May, as distinct from October, and from early spring, troughs in the subtropical westerlies become stationary across the Bay of Bengal often enough to affect the mean resultant circulation. They may also connect with West China troughs to the north. At 500 mb (Fig. 5.35) in October, a ridge spans the Bay of Bengal at 19N, whereas in May a trough stretches from Ceylon to East Pakistan. The greater baroclinity in May is also evident at 200 mb; the subtropical ridge lies 3° farther south than in October.

The near-equatorial trough intensifies and moves north sufficiently between April and May (Fig. 5.18) to be activated by upper divergence east of a trough in the subtropical westerlies. The first disturbance so to develop starts the summer monsoon *rains* of Burma and, over northeastern India, expands and intensifies to Bengal the *rains* previously confined to Assam. Upper *convergence* west of the trough keeps weather over central India drier in May than in October although there is little difference in cloudiness.

5.8.4.3 *Movement of Tropical Disturbances in May and October.* The trough in the subtropical westerlies also affects disturbance tracks; a much greater proportion recurve toward the northeast in May than in October, over both the Bay of Bengal and the South China Sea (Fig. 5.35).* Consequently rainfall over eastern peninsular India, eastern Vietnam, and even the Paracels, amounts to only 15–20% of the October values (Figs. 5.32, 5.46).

A disturbance moving up the Bay of Bengal east of a combined subtropical trough and West China trough injects much extra moisture into the middle-tropospheric westerlies flowing across Yunnan to southern China (Section 5.8.4.1). Thus a burst of Burma and Bengal monsoon *rains* accompanies increased spring *rains* over southern China.

The Burma *rains* (which come from subtropical as well as tropical cyclones) begin as early as late April or as late as early June (Huke, 1965). When they are early, the West China trough is vigorous and the South China spring *rains* are also copious.

---

* However, over the Arabian Sea tropical cyclones move west northwest in both months; on an average, the Arabian coast is affected only once every 3 yr (Pedgley, 1969b).

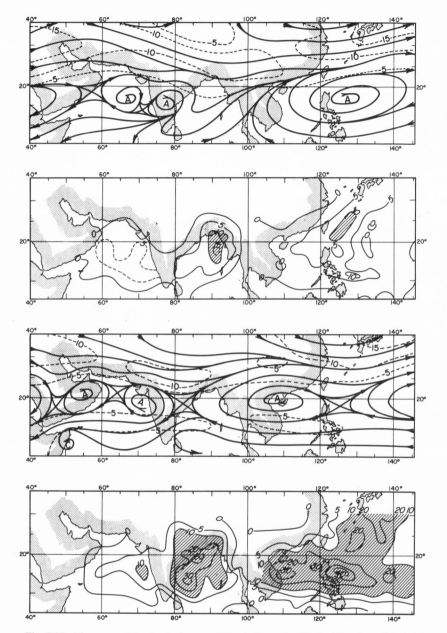

Fig. 5.35. Mean monthly circulations at 500 mb. Isotachs (dashed lines) are labeled in meters per second. Number of recorded tropical cyclone tracks passing through each $2\frac{1}{2}°$ square for a 70-yr period (thin, full lines) (Chin, 1958; Ananthakrishnan, 1964). (top pair) May; (bottom pair) October.

5.8.5   SUMMER (JUNE TO SEPTEMBER)

Summer marks the period of heat trough predominance. The near-equatorial trough is seldom noticeable except during " breaks " in the monsoon *rains* (Section 5.8.7.5), and westerly or southwesterly surface winds prevail everywhere south of 20–25N.

Nevertheless the season is not homogeneous. It can be conveniently divided in two—advance of summer (June to mid-July), and height of summer and its wane (mid-July to September).

5.8.6   ADVANCE OF SUMMER (JUNE TO MID-JULY)

In local day-to-day detail, the advance of summer over monsoon Asia is bewilderingly complex. Usual preconditions such as heat-low intensification and subtropical jet weakening, may develop long before or coincidentally with, or even subsequent to, rain onset, and then once the rain starts, it in turn may hasten the further advance of summer. Nevertheless, *major* changes follow the same sequence each year. In a particular year they may occur on dates differing by two or three weeks from the average dates presented in Fig. 5.17 (Ramaswamy, 1969). Nevertheless, events dependent on a single major change, such as onsets of rain at Delhi and of the midsummer dry spell at Hong Kong tend to maintain the same temporal relation as shown by the median dates of these events.

During June, tropical and subtropical cyclones develop in the Bay of Bengal and more rarely in the South China Sea (Ramage, 1959) (Section 5.8.7.6). In the fresh, convergent southwesterlies south of the centers rainfall is extensive and often heavy.

Table 5.9

Annual variation of 300-mb mean resultant zonal winds (m sec^{-1}, west positive) north of Tibet and south of the Himalayas

	Jan.	Feb.	Mar.	Apr.	May	June	July	Aug.	Sept.	Oct.	Nov.	Dec.
Kuche 41°45′N, 83°04′E	20	16	20	19	17	17	25	25	22	16	17	21
New Delhi 28°35′N, 77°12′E	34	34	26	21	19	9	−1	−1	6	21	28	33

5.8.6.1   *Early June.* Sensible heating (Murakami, 1958) and condensation heating from numerous thunderstorms over southeastern Tibet (Flohn, 1968) rapidly weaken the subtropical jet south of the Himalayas, in comparison to the polar jet along the northern edge of the plateau (Table 5.9).

Moist surface air from the Bay of Bengal may reach as far as Kashmir. Then, a rare late-season western disturbance entering the region triggers widespread thunderstorms over northwestern India (Mooley, 1957). In such a year, monsoon rain starts earlier there than in the south, confusing those who believe in a northwestward advance over the subcontinent.

The fitful and diminishing subtropical jet in association with the West

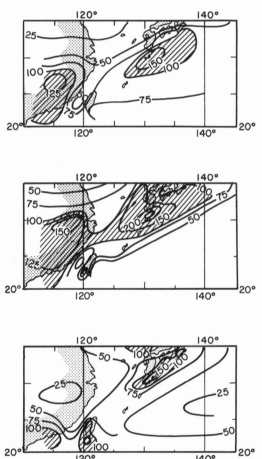

Fig. 5.36. Long-term mean rainfalls (mm) for 10-day periods (Yoshino, 1966): (top) 21 through 30 May; (middle) 20 through 29 June (Mei-Yü); (bottom) 10 through 19 July (post Mei-Yü).

China and subtropical westerly troughs continues as in May to enhance rain from tropical and subtropical cyclones over Burma and northeastern India (Ramaswamy, 1969) and from disturbances on the polar front over North Vietnam and southern China. Rainfall in these regions originates primarily from moisture evaporated by fresh winds from the Bay of Bengal. The South China Sea contributes less (Figs. 5.36 and 5.37). The polar jet is already stronger than the subtropical jet. By facilitating southward incursion of relatively cold air to the east of Tibet, it further intensifies the polar front (Somervell and Adler, 1970).

Weakening of the Himalayan subtropical jet removes a hindrance to large-scale upward motion to the south (Section 5.1.2), and according to Yin (1949) precedes onset of monsoon rain. However, Ramamurthi and Jambunathan (1967) found for southwestern peninsular India that the subtropical

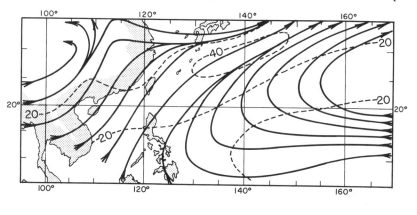

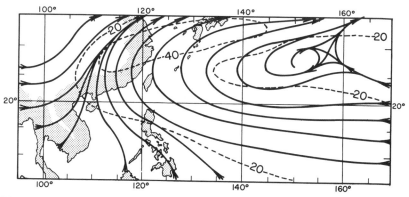

Fig. 5.37. Ten-day mean integrated (surface to 500 mb) water vapor transfer shown by streamlines (direction) and isopleths labeled in units of millibar meters per second (Murakami, 1959b): (top) 11 through 20 June 1957; (bottom) 1 through 10 July 1957.

westerlies remained unchanged or were increased just prior to the rain or even after it had started. The only firm prerequisite seemed to be formation of a middle-tropospheric trough over southern India a few days before. Then development of a tropical depression or subtropical cyclone in the trough brought the rain. Presumably in the far south rain is seldom controlled by events near the Himalayas.

The first falls, in which the upper troposphere retains the relative dryness and instability of spring (Fig. 4.2), are frequently from thunderstorms.

5.8.6.2 *Mid-June*. The subtropical jet has disappeared, the West China trough is no longer significant, and a lower-tropospheric wedge from the Pacific high often extends across southern China.

(*i*) *Development of the Mei-Yü over China.* The intensified polar jet, sweeping south around northeastern Tibet, activates the polar front which has shifted north to the Yangtze Valley (Dao *et al.*, 1958; Zou *et al.*, 1964) as rainfall over southern China begins to diminish (Fig. 5.36). Usually a considerable air mass discontinuity across the front and very slow-moving disturbances ensure copious Mei-Yü or plum *rains*.

According to Chu (1962), during the Mei-Yü the Okhotsk high blocks rapid eastward progression of disturbances on the polar front, while maintaining significant air mass contrast across the front.

Asakura (1968) reported that when the subtropical jet south of the Himalayas weakens, and the polar jet north of the Himalayas strengthens, a blocking high develops over the Sea of Okhotsk. Thus the early summer rain over India tends to fluctuate in unison with the Mei-Yü. From a simple numerical model of the early summer circulation, Asakura deduced that the Okhotsk blocking high results from the " union of a large heat source near India and a cold source near the Okhotsk Sea." The connection is statistically apparent; between 21 May and 19 June, 5-day mean surface-pressure anomalies over northwestern India and over the Sea of Okhotsk are negatively correlated ($r = -0.56$, $n = 30$).

From this one obtains a glimpse of the tangle of southern and eastern Asia interactions which must be unravelled before the weather can be properly understood (Section 7.3).

The wider spacing of average rain-onset isochrones between southern and central China as compared with these to north or south (Fig. 5.17) might reflect the jump from subtropical to polar jet domination. The dividing zone between the spring and early summer polar front *rains* of southern China and the plum *rains* of central China lies between 26 and 27N (Chu, 1962).

During the course of the Mei-Yü, the prime moisture source for the *rains* shifts from the Bay of Bengal and the South China Sea (Fig. 5.37) to the western Pacific. Transport is increasingly effected by southeasterlies which turn anticyclonically to southwesterlies across southeastern China. Over Kweichow, Chao (1965) reported the "southwest" monsoon is usually replaced by the "southeast" monsoon early in July. In both regimes the strongest moisture convergence occurs in the polar front zone.

Murakami (1959a) computed the energy budget of the Mei-Yü between the surface and 500 mb over the period 21 to 30 June 1957. He resolved the field of motion into the time-averaged motion and the superimposed large-scale eddies and found that both generated kinetic energy surpluses in the region. The net outward flux of meanmotion kinetic energy, which was an order of magnitude greater than the net outward flux of eddy kinetic energy, amounted to 4.96×10^{19} ergs sec^{-1} eastward to the south of Japan and 3.50×10^{19} ergs sec^{-1} upward through the 500-mb level. The vehicle was jetlike middle-tropospheric west-southwest winds.

In a similar treatment of available potential energy [that part of the potential energy available for transformation into kinetic energy under any adiabatic redistribution of mass (Lorenz, 1955)], Murakami computed a net influx to the Mei-Yü region of mean available potential energy of 5.11×10^{19} ergs sec^{-1} predominantly from south and southwest (Fig. 5.37), and a net influx of eddy available potential energy of 5.18×10^{19} ergs sec^{-1} predominantly from the Sea of Okhotsk region. Along the active polar front zone, warm air rising to the south and cold air sinking to the north transformed the available potential energy into an exportable surplus of kinetic energy. One would expect then that Mei-Yü intensity would be reflected in the vigor of northern Pacific circulations.

(*ii*) *Development of the Tibetan high.* Over India the Tibetan high is increasingly evident, and heavy, but not sustained monsoon rain becomes more common over central and western India through the agency of monsoon depressions and subtropical cyclones. As moisture spreads deeply through the troposphere, lapse rates decrease, thunderstorms lessen, and *rains* predominate (Fig. 4.2).

5.8.6.3   *First Half of July.* In the upper troposphere, westerlies disappear south of the Himalayas. For the rest of the summer they rarely reappear and then generally only when the monsoon rain "breaks" (Section 5.8.7.5). The Tibetan high and divergent easterlies to the south facilitate development of disturbances and are in turn intensified by latent heat of condensation released in these disturbances (Section 3.2.3). Monsoon rain spreads over south Asia east of 70E.

Over China, in the upper troposphere, as the Tibetan high develops, the polar jet shifts northward. With the Okhotsk high weakening, the polar front also weakens and shifts north and the Yangtse Valley Mei-Yü is over (Dao *et al.*, 1958; Zou *et al.*, 1964) (Fig. 5.36). A *rains* regime, with precipitation on more than half the days, is replaced by a *showers* regime. Thunderstorm frequency and rain intensity both increase (Murakami, 1959b; Chu, 1962), although precipitation is now recorded on only one-third of the days (Fig. 5.43).

Southern China, now well out of range of the polar front, experiences a dry spell (Ramage, 1952b) (Fig. 5.38). Median dates for the start of the dry spell at Hong Kong and for the onset of monsoon rain at Delhi (Bhullar, 1952) are 5 and 3 July, respectively, and the correlation coefficient of the two events is 0.71 ($n = 37$). This is persuasive evidence of the widespread effects produced by major circulation changes.

5.8.7   HEIGHT OF SUMMER AND ITS WANE (MID-JULY TO SEPTEMBER)

The fully established Asian summer monsoon extends from northeastern Africa to the Philippines and from 30N to 10S, displacing the Hadley cell to north of the Himalayas.

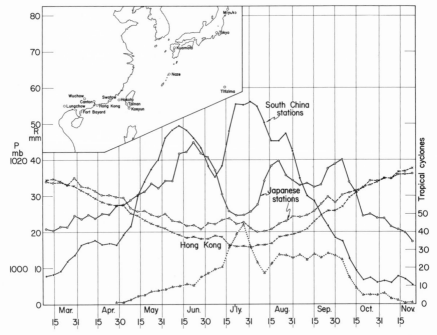

Fig. 5.38. Five-day means of rainfall (full lines) for nine stations in southern China and five stations in and south of Japan, pressure (dashed lines) for Hong Kong and the five Japanese stations, and 50-yr tropical cyclone frequencies (dotted line) for southern China.

Circulation patterns do not change during September, but the summer monsoon steadily weakens. Pressures rise in the heat trough, winds in lower and upper troposphere decrease, and rainfall diminishes everywhere except over eastern peninsular India and eastern Vietnam where tropical disturbances mark the beginning of the rise to the autumn maximum (Section 5.8.1.2).

5.8.7.1 *Large-Scale Monsoon Circulation.* The *average* Asian summer monsoon circulates in two intersecting modes. In the zonal mode westerlies predominate between 1.5 and 3 km and easterlies between 8 and 10 km. In the transequatorial mode, surface southwesterlies are overlain by eastnortheast winds between 12 km and the tropopause. In both modes, rising motion predominates east of 70E and sinking motion west of 70E (Section 5.1.2).

The 10-day periods 21 to 30 July 1957 and 8 to 18 July 1958, described by Dao *et al.* (1962), typify the two modes. In the former, zonal circulations prevailed in both hemispheres; in the latter, a meridional circulation extended across the equator and typhoons were active over the China Seas.

The two modes share an upward branch, generally enclosed within the area 70–120E and 10–25N. Their downward branches however are separate. In the zonal mode, sinking occurs west of 70E over the Arabian Sea and the Northern Hemisphere deserts. In the transequatorial mode sinking, though it occurs south of 5N, is probably concentrated north of the South Indian Ocean subtropical jet (Section 7.1.2).

At the Seychelles (4°37'S, 55°27'E) there is evidence of subsidence accompanying the transequatorial mode; 700-mb humidities are lowest with 200-mb winds from between 48 and 77°, and highest with 200-mb winds from between 132 and 311° (Wright and Ebdon, 1968).

Thus not only meridional temperature gradients, but also zonal temperature gradients (Fig. 2.4), control the monsoon circulation and its fluctuations.

Heat of condensation added to the rising air is transported to the heat sink where the air subsides. As in the winter monsoon of eastern Asia (Section 5.8.2.3), any mechanism which increases the temperature difference between source and sink *will accelerate the circulation*—a typhoon over the South China Sea, a monsoon depression over the Bay of Bengal, or even a large-amplitude trough in the Southern Hemisphere polar westerlies, for example:

(1) Freshening of low-level southwesterlies over peninsular India and the central Bay of Bengal, accompanied by heavier rain in the eastern Bay, might result from distant events. In turn, this freshening, by increasing cyclonic vorticity to the north, might lead to the development of a monsoon depression in the preexisting trough (Section 3.4.2). Similar surface surges sometimes precede appearance of a typhoon over the western Pacific or the South China Sea, while surges at the 3-km level may herald subtropical cyclogenesis.

(2) Between 21 and 23 July 1966 surface pressure dropped 2–8 mb every-where over monsoon Asia (Sadler *et al.*, 1968). Widespread and often heavy rain fell south of the monsoon trough where surges had set in. Within two days, development of a typhoon (ORA) in the South China Sea and a low-pressure system in the Bay of Bengal further intensified the rain. Similar extensive pressure falls over India were first described by Eliot (1895), while falls over the Caribbean and the Indian and central Pacific oceans, apparently unconnected to synoptic events, were described by Frolow (1951) and Palmer and Olmstede (1956).

5.8.7.2 *A Numerical Model of the Monsoon along 80E*. Murakami *et al.* (1970) applied the equations of motion to an eight-layered atmosphere to develop a two-dimensional numerical model of the summer monsoon extend-ing along 80E from the equator to the North Pole. Neither disturbances nor transequatorial flow were permitted. Effects of radiational and condensation heating were allowed for; within the turbulence layer, vertical fluxes of momentum, heat, and water vapor were assumed to be constant and were computed using empirical formulas. The investigators started their compu-tations with a completely calm and dry atmosphere possessing the "standard atmosphere" vertical temperature distribution. In their latest experiment they included evaporation at the earth's surface, condensation in the atmo-sphere, and mountain effect. Sea-surface temperature was kept constant at 27C, and the following assumptions were made: land albedo, 15%; surface relative humidity over the continent, 60%; surface relative humidity over the ocean, 100%; and condensation relative humidity, 100%. Integration carried out over a period of 80 days using 10-min time steps, determined a zonal wind profile (Fig. 5.39). Comparison with the long-term mean for 73E shown in Plate III reveals only slight discrepancies. Values of speed maxima agree. This is particularly significant since in earlier experiments which did not include the mountain effect, the westerly maximum was computed to be less than 5 m sec^{-1} and the easterly maximum less than 10 m sec^{-1}, comparable to averages along cross sections *east* of the Himalayas. Thus the model strikingly confirms the potent effect of Himalaya–Tibet on the circulation (Section 5.1.2).

5.8.7.3 *Monsoon Subdivisions*. Although together they comprise a single system, three dissimilar regional circulations exist and will be treated sep-arately:

(1) *West of 70E:* middle-tropospheric subsidence, large vertical wind shear, and fine weather.

(2) *Between 70E and 105E:* large-scale upward motion, large vertical wind shear, monsoon depressions, subtropical cyclones, and frequent *rains*.

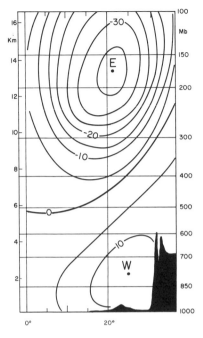

Fig. 5.39. Computed zonal wind in meters per second for July along 80E obtained after an 80-day time integration (Murakami *et al.*, 1970).

(3) *East of 105E:* intermittent upward motion, small vertical wind shear, typhoons, subtropical cyclones, and changeable weather.

5.8.7.4 *West of 70E.* Above the heat trough and over most of the Arabian Sea, northeastern Africa, and Arabia, the upper-tropospheric easterlies converge and subsidence prevails through the middle troposphere (Flohn, 1964c) (Section 5.1.2).

The combination of large vertical wind shear and subsidence prevents tropical cyclones developing (Ramage, 1959) (Section 3.3.1). Broken middle and high clouds, which are often observed over the eastern Arabian Sea, probably originate in rain systems farther east. Showers are rare and cumulus humilis predominates (Bunker and Chaffee, 1970). Persistent haze stems from two causes. Over the West Arabian Sea dust blown off northeast Africa is present through a deep layer. Over the East Arabian Sea haze occurs when —environmental relative humidity exceeding 80%—water condenses on hygroscopic sea-salt particles previously whipped up from the sea surface by strong winds (Bunker and Chaffee, 1970).

(*i*) *Air–Sea interaction over the Arabian Sea.* Upwelling, which began in May off Somalia and Arabia (Section 5.8.4.2), is intense during July and

August, and, through feedback, has so increased the pressure gradient between sea and land that the southwesterlies are the strongest and most-
sustained surface-wind system on earth (McDonald, 1938). Where upwelling
occurs, summer sea-surface temperatures are lower than in any other season
(Wooster *et al.*, 1967). However, even beyond the reach of upwelling, strong
winds sufficiently cool the water by evaporation to produce a secondary sea-
surface temperature minimum (Colon, 1964) (Fig. 5.7). The evaporated
water, condensing farther east in subtropical cyclones and monsoon depressions, contributes to the energy of these systems.

(*ii*) *Example: Upwelling and weather off Somalia, August 1964* (Ramage,
1968a). The research vessel *Discovery* was in the area during the summer of
1964. Observations made from the ship, when combined with an understanding of the tropospheric circulation over the Arabian Sea, help to explain
summer absence of fog. Between 16 and 21 August 1964, *Discovery* sailed
northward along about 50E from 6.5 to 12.3N (Fig. 5.40). making numerous

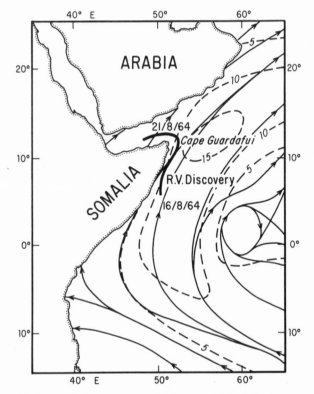

Fig. 5.40. Kinematic analysis of surface winds (m sec $^{-1}$) for 0600 GMT, 20 August 1964.
Discovery's route is shown by heavy line (From Ramage, 1968a).

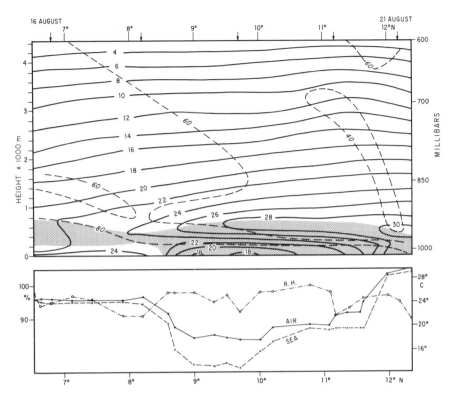

Fig. 5.41. Analysis of measurements made along about 50E by *R. V. Discovery* between 16 and 21 August, 1964. (Top) Isotherms (°C, full line), isohumes (percent, dashed line), and inversion layer (stippled). Arrows denote locations of radiosonde ascents. (Bottom) Air temperature at deck level (full line), sea-surface temperature (dashed line), and relative humidity at deck level (dot–dashed) (From Ramage, 1968a).

sea-surface and air-temperature measurements as well as five aerological soundings (Fig. 5.41). The synoptic situation, which changed little during this period, is illustrated by the surface analysis for 0600 GMT 20 August (Fig. 5.40).

From 16 to 21 August, *Discovery* never recorded more than 10% low cloud nor visibility below 10 km. Winds blew persistently from between south and southwest at from 10 to 15 m sec^{-1}, diminishing only after the ship sailed to the lee of Cape Guardafui.

At the outset, air and sea temperatures were the same and a weak inversion extending from about 200 to 800 m inhibited low-cloud development. Then when the ship encountered cold water near 8.5N, the inversion extended to the surface, being intensified both from below and above to greater than 10C. Although the relative humidity of the surface air exceeded 90%, no fog developed, despite the fact that the air moving over the cooling surface had possessed, but 24 hr before, a dew point 9C higher than the temperature of the upwelled water.

Discovery's soundings reveal a rapid decrease in relative humidity accompanied by a slight increase in mixing ratio with height in the inversion, thus confirming the existence of a downward moisture flux (Roll, 1965). The strong surface winds no doubt facilitated downward transport of heat and moisture, despite the great stability of the air.

At the surface, the effect of upwelling cold water was to increase the downwind pressure gradient and so to accelerate the winds. The consequent divergence (see Fig. 3.37), by bringing the subsidence inversion down to the surface, ensured a supply of dry air adequate to prevent fog from forming (Section 5.8.3.2).

Since summer fog never develops off Somalia (Meteorological Office, 1949), a surface subsidence inversion must be a season-long phenomenon.

5.8.7.5 *Between 70E and 105E.* In Sections 3.4 and 3.5 I described development, intensification, and decay of monsoon depressions and subtropical cyclones. These, together with the monsoon trough over northern and northeastern India and the strong but fluctuating surface southwesterlies and uppertropospheric easterlies, comprise the main components of the Indian summer monsoon.

(*i*) *Monsoon trough.* The trough is so sharply defined on mean charts (Fig. 5.42) (Ananthakrishnan and Rao, 1964) that it is most likely topographically anchored. Since it lies parallel to and about 450 km southwest of the Himalayas, they probably constitute the anchor. Anabatic winds on the southern slopes of the Himalayas are part of the large-scale summer circulation (Flohn, 1968). Though strongest in the afternoon, they prevail throughout the 24 hr of the day. They coincide with the northern rainfall maximum of Fig. 5.42 and presumably form part of a local vertical circulation in which air returns southwestward in the middle troposphere and tends to sink over the northern plains. That subsidence warming diminishes surface pressures, inhibits rain, and, by increasing the middle tropospheric lapse rate, favors a *showers* regime is confirmed by Fig. 5.42. The monsoon trough roughly coincides with a relative rainfall minimum and a relative maximum in thunderstorm frequency. A mechanism operating 24 hr a day would also explain why the monsoon trough displays little diurnal intensity variation (Ananthakrishnan and Rao, 1964). Monsoon depressions usually move parallel to but south of the monsoon trough. This is consistent with the observation that heavy rains are confined to a strip on the left of the track (Section 3.4.1); but can we confidently apportion cause and effect?

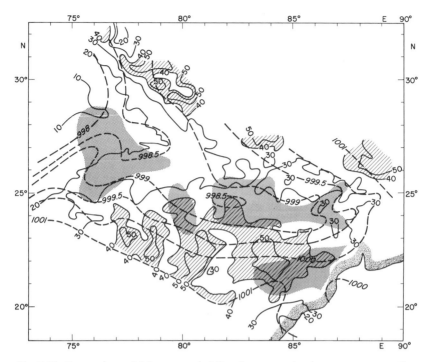

Fig. 5.42. Comparison of July mean rainfall and mean sea-level pressure over northern India. Full lines are isohyets labeled in centimeters and are not shown above 50 cm; dashed lines are isobars labeled in millibars. Stippling denotes areas with an average of more than 10 thunderstorm days in the month (from India Meteorological Department, 1953, 1962).

(ii) Monsoon rains. Sometimes copious monsoon *rains* over central and western India last for several weeks.

The *rains* usually set in when lower-tropospheric southwesterlies freshening over southern India and the Bay of Bengal trigger a monsoon depression at the head of the Bay (Section 3.4.2). In turn, subtropical cyclones (Section 3.5.5), lesser low-pressure systems, or monsoon depressions develop, maintaining a vigorous monsoon circulation with very little let up in the *rains.*

Just as in other parts of the monsoon area (Sections 5.5.2, 5.8.2.1, and 5.9.2), orography is not the *prime determinant* of rainfall distribution. Less rain falls on the Kirthar Range (Fig. 3.19), directly exposed to the monsoon surface winds, than over the Deccan Plateau of central India, lying in the rain shadow of the Western Ghats, because large-scale subsidence prevails over the former and large-scale rising motion over the latter.

However, orography determines the *detailed pattern* of rain when the monsoon is active. Heaviest falls occur on windward slopes of the Western

Fig. 5.43. Clouds photographed from ESSA 5 between 0530 and 1100 GMT, 25 July 1967 (Orbit Nos. 1214–1216).

Ghats and the Khasi Hills (Section 4.2 2.1), while to the lee of the Ghats rain diminishes sharply and takes on a *showers* character (Section 4.1.2).

(*iii*) *Example: A day of extensive rain, 25 July 1967* (Fig. 5.43). Three separate low-pressure systems contributed to the activity, subtropical cyclones at 23N, 71E (strong) and at 21N, 106E (weak), and a monsoon depression at 20.5N, 87.5E. A possible fourth low was centered near 10N, 115E. To the west–southwest of each of these systems a lower-tropospheric westerly maximum was causing significant downstream convergence where clouds were thickest, while at 200-mb divergent flow coincided with the cloud masses. At Bangkok (Fig. 5.44) a deep moist layer and relative stability were typical of a *rains* regime.

A weakly defined upper-tropospheric ridge lay south of its normal latitude and winds were light and variable over northern India. Blobs of intensely

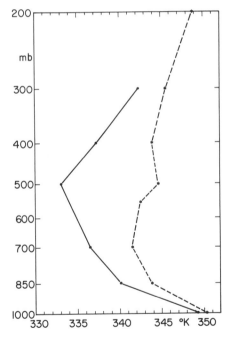

Fig. 5.44. Vertical profiles of average virtual equivalent potential temperature (θ_{ve}) in degrees Kelvin measured at 0000 GMT at Bangkok (13°55′N, 100°36′E) for 5 through 6 July 1967 (full line) and 24 through 26 July 1967 (dashed line) (Harris *et al.*, 1970). Ten stations in the Bangkok area averaged 0.3 mm of rain in 24 hr on the 5th and 6th, and 6.0 mm between the 24th and 26th. Rain was recorded during one percent of 3-hourly or 6-hourly intervals on the 5th and 6th, and during 32% of the intervals between the 24th and 26th.

reflecting clouds lay along the well-marked monsoon trough near 25N (Fig. 5.42); surface observations confirmed that some were thunderstorms. Over China the Mei-Yü had given way to widespread thunderstorms (Section 5.8.6.3, Fig. 5.36).

Low cloud streaks over the central Arabian Sea show the anticyclonic turning typical of fine weather.

A rough idea of the intensity of the monsoon circulation (Section 7.1.2) can be obtained from the 2–3-km winds at Port Blair (11°40′N, 92°43′E) (230°18 m sec^{-1}) and the 12–14-km winds at Gan (0°41′S, 73°09′E) (70°41 m sec^{-1}).

(*iv*) *"Breaks" in the monsoon rain.* A few times every summer, a "break" of several days or more occurs (Fig. 2.8) as *rains* shift from central India to the Himalayan foothills, the monsoon trough shifts an average of 250 km in the same direction (Ramaswamy, 1969), and weather deteriorates over southern India and Ceylon.

Pisharoty and Desai (1956), Ramaswamy (1958, 1962, 1967), and Parthasarathy (1960) thought that a "break" develops when the upper-tropospheric ridge (including the Tibetan high) along 30N weakens and a large-amplitude trough or a series of troughs in the polar westerlies or western disturbances protrude south of the Himalayas—a low westerly-index situation (Ramaswamy, 1962). In the Himalayan foothills rain increases at the approach of each western disturbance (Chakravortty and Basu, 1957) and diminishes as it passes, but the bad weather does not extend into the plains. Ramaswamy has intensively studied this problem. Most recently (1969), he said:

> "We do not suggest that the westerly jet stream itself moves into India during break-situations. We merely state that during the break-conditions, pronounced low-index circulation in the westerlies to the north of India brings the [westerly jet] equatorward and nearer to the [easterly jet] causing dynamical interaction between the two systems."

Koteswaram (1950) on the other hand, associated "breaks" with weak mid-tropospheric lows moving slowly westward across and causing weather to deteriorate over southern India and Ceylon. Dixit and Jones (1965) ascribed a "break" to extension of a mid-tropospheric ridge from Indo-china across central India.

Dixit and Jones (1965) and Ramamurthi *et al.* (1967) observed that during "breaks" the heat trough and lower-tropospheric southwesterlies and upper-tropospheric easterlies diminished (Section 3.2.3).

Since the rainfall distribution and other features superficially resemble May or October normals (Section 5.8.4.2), it is tempting to conclude that the circulation during a "break" merely takes on a transition season character, with rising motion at the subtropical jet and the near-equatorial convergence and sinking between.

This would imply a breakdown of the vertical monsoon circulation with exchange between Northern and Southern Hemispheres largely confined to near-equatorial latitudes (Koteswaram, 1950), no more than a remnant monsoon cell over central India, and an invigorated Hadley cell extending northward from the Himalayas (Ramamurthi et al., 1967).

Nevertheless, in the well-documented studies by Dixit and Jones (1965) and by Ramamurthi et al. (1967), the trough in the polar westerlies criterion was not often satisfied, the Tibetan high usually shifted only imperceptibly, and the westerly and easterly jets moved slightly *apart* as a " break " developed. Although Koteswaram's criterion seems usually to be satisfied, it was not in Ramaswamy's (1967) case study.

The apparent inconsistencies may arise from difficulties in unequivocally defining a " break," a particularly tricky task if half of central India is dry and the other half is wet.

However, the problem is likely to be more fundamental and even less tractable. Although some breaks apparently result from southward extension of troughs in the upper-tropospheric westerlies, others may reflect changes in the zonal or transequatorial mode of the vertical monsoon circulation (Section 5.8.7.1). Most probably, every break has its own peculiar combination of antecedents which is unlikely to be revealed by synoptic climatological studies.

On an average, breaks occur more frequently between 10 and 20 August than before or after. This is reflected in a secondary minimum in 5-day mean rainfalls at Indian stations north of 16N (Ananthakrishnan and Pathan, 1970).

(v) *Example: A break in monsoon rains, 6 July 1967* (Fig. 5.45). There were scattered showers and some thunderstorms beneath a 500-mb ridge which extended across southern China, the Gulf of Tonkin, and Thailand to India. Cirrus streamed westward from cumulonimbus tops penetrating the upper-tropospheric easterlies.

The midsummer dry spell had begun over southern China (Section 5.8.6.3) and *rains* were confined to Japan, central China (Mei-Yü, Fig. 5.36), and the foothills of the eastern Himalayas. The monsoon trough lay close to the mountains over India. A mid-tropospheric low which had apparently developed over southwestern peninsular India on the 4th was moving slowly westward (Koteswaram, 1950). Rainfall over the Laccadive Islands ranged from 3 to 17 cm.

Relative dryness and instability at Bangkok (Fig. 5.44) typified a *showers* regime. Stretching southwestward from typhoon BILLIE, centered east of Taiwan, a *rains* zone covered most of the Philippines.

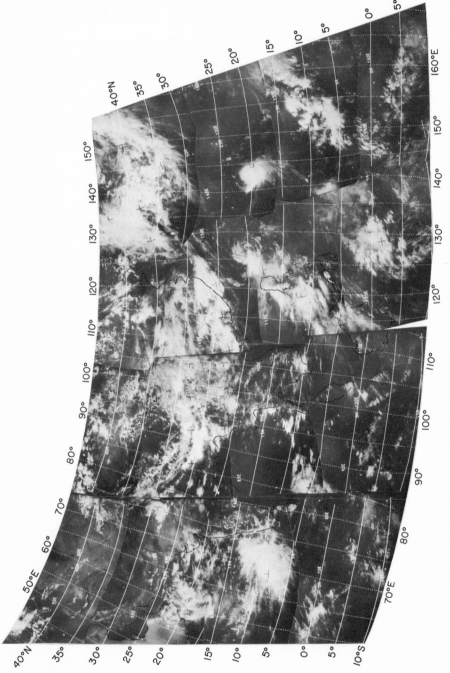

Fig. 5.45. Clouds photographed from ESSA 5 between 0450 and 1050 GMT, 6 July 1967 (Orbit Nos. 972–975).

Further evidence of a weakened monsoon circulation (Section 7.1.2) comes from the 2–3-km winds at Port Blair (200°, 6 m sec^{-1}) and the 12–14-km winds at Gan (90°, 28 m sec^{-1}).

5.8.7.6 *East of 105E: The Typhoon Season.* At the end of the Mei-Yü, and as the weakened polar front moves to the north, the circulation east of 105E becomes an extension of the circulation west of 105E. However, although mean streamlines reveal little difference between the two regions, absence of Himalaya–Tibet east of 105E results in a sharp discontinuity in circulation *intensity* along that meridian.

(*i*) *Monsoon trough.* During the early July dry spell over southern China (Fig. 5.38), heating facilitates eastward extension of the monsoon trough which joins with the near-equatorial trough across the Philippines.

On mean charts, the monsoon trough can barely be detected as a broad zone between southwesterlies to the south and southeasterlies to the north. Absence of the protecting, intensifying, and anchoring Himalayas accounts for the relative climatological weakness of the trough and for the fact that it follows the sun northward, attaining its highest average latitude of 30N in the first half of August. Thereafter it shifts south, ceasing to be a monsoon trough during September and giving way to a dominant near-equatorial trough across the relatively warm South China Sea in October (Section 5.8.1.2).

On synoptic charts the monsoon trough is more sharply defined than in the mean. It is the birthplace of disturbances. As in the Indian monsoon trough and in the near-equatorial trough, a secondary rainfall minimum coincides with the trough axis in contrast to rainfall distribution about the polar front (Fig. 5.34).

(*ii*) *Typhoons.** (1) Development. The monsoon trough provides a surface focus for typhoon development which is further facilitated by a tendency toward large-scale rising motion (Section 5.1.2) and little vertical wind shear (Plate III).

During July, increasing numbers of typhoons develop over the western Pacific and the China Seas.

In the upper troposphere, at the end of the Mei-Yü as the subtropical ridge moves north, a trough appears between 20 and 25N near 165E (Yoshino, 1967). This "mid-Pacific trough" (MPT) (Ramage, 1959) stretches east–northeast from the Marshall Islands into middle latitudes and persists through October, separating the subtropical ridge from another ridge along 10N

---

* In this category I include less intense warm-cored tropical depressions.

(Wiederanders, 1961). The MPT is a potent typhoon producer, apparently through downstream energy dispersion (Ramage, 1959) or more directly when cyclonic vorticity spreads downward to the surface from a vigorous cyclone in the trough (Sadler, 1967). Over the western Pacific, the Hadley cell has been displaced by the summer monsoon, whereas it persists in its normal location over the eastern Pacific. Conceivably, adjustment between these disparate circulations produces the MPT and the ridge to the south. Without the Asian summer monsoon, the MPT would be no stronger than the Mid-Atlantic trough (La Seur, 1967), and without the monsoon, surface wind surges which often precede and may trigger typhoons (Section 5.8.7.1) would be much weaker. Surely then, juxtaposition of the world's greatest monsoon system and most active typhoon-generating region cannot be fortuitous.

(2) Movement. Upper-tropospheric easterlies "steer" many typhoons toward southern China, where typhoon frequency and rainfall sharply increase after 10 July (Fig. 5.38).

On an average, as the monsoon trough shifts north and then south with the sun, so do the typhoon tracks (Table 5.10), southern China experiencing a

Table 5.10
Average number of East Asian typhoon landfalls in 10 yr[a]

Coastal zone	Jan.–Mar.	Apr.	May	June	July	Aug.	Sept.	Oct.	Nov.	Dec.
China, north of 28N	0	0	0	0.2	2.3	5.0	1.4	0	0	0
China, between 28N and 23.5N	0	0	0	2.3	7.3	10.4	6.0	1.7	0	0
China, south of 23.5N	0	0	1.4	1.4	4.1	4.3	6.0	3.2	0.6	0.2
Vietnam	0	0.6	0.2	1.7	3.1	1.7	5.2	7.1	3.3	1.2

[a] From Kao (1951).

secondary frequency minimum and decreasing rainfall in early August (Fig. 5.38). In early June typhoons may cross Taiwan on recurring tracks but in mid-August west-northwest or northwest tracks predominate (Chin, 1958).

Return of the weakening monsoon trough to southern China near the end of August shows in the mean merely as a temporary decrease in the rate of pressure rise and as a slight secondary rainfall peak (Fig. 5.38).

(*iii*) *Subtropical cyclones.* Subtropical cyclones develop over the South China Sea and Indochina during summer. They are rarer and less intense than their counterparts over the Northeast Arabian Sea and some come to resemble monsoon depressions. Accompanying rains extend in a broad swath southwestward from each center (Fig. 5.43). Over the Mekong Valley, subtropical cyclones account for 25% of the annual rainfall (Rainbird, 1968). Broadly speaking, subtropical cyclone life cycles follow one of three courses:

(1) Over northern Thailand or Indochina, a low becomes cut off from a large-amplitude middle-tropospheric trough east of Tibet. The sequence bears some resemblance to development of winter Kona storms near Hawaii (Simpson, 1952). After a few days the cyclone weakens.

(2) One to three days prior to development, cyclonic westerlies set in across southern Indochina and the western part of the South China Sea. The winds can reach 20 m sec^{-1} at 3 km but are light at the surface. Occasionally, over the South China Sea a surface low may form on the eastern edge of the middle-tropospheric vorticity center, move westward, and intensify, reaching maximum strength (gale-force winds) when it comes beneath the upper vorticity center. The system remains stationary for two or three days before finally drifting inland over South China or Indochina and weakening (Ramage, 1959).

Monsoon winds freshen before a cyclone forms (Section 5.8.7.1) and then increase further in response to the cyclone-generated pressure gradient. However, the strongest winds set in at a higher level prior to subtropical cyclone development than prior to or accompanying typhoon development.

(3) As a vigorous typhoon moves inland over South China or Tonkin and *rapidly* fills, a midtropospheric cyclonic vortex may develop over the South China Sea or Vietnam within the deep westerlies which just previously flowed strongly toward the typhoon. The new circulation then intensifies into a subtropical cyclone as in paragraph (2). The sequence fits an hypothesis advanced by Bergeron (1954). According to him, rapid pressure rise in a filling typhoon must be compensated by pressure fall elsewhere, possibly through a wavelike, energy dispersion (Yeh, 1949) mechanism.

(*iv*) *Weather near the monsoon trough.* Typhoons and weaker tropical disturbances account for most of the bad weather. Their most-traveled tracks do not coincide with areas of maximum average cloudiness which lie farther south where southwesterlies prevail (Plate III).

(*v*) *Weather south of the monsoon trough.* Convergence, ahead of surges in the southwesterlies associated with and sometimes preceding typhoon or subtropical cyclone development to the north (Section 5.8.7.1), accounts for intensification of summer *rains* south of the monsoon trough (Fig. 5.43). Orography figures importantly in modifying distribution, intensity, and even the character of the rain (Section 4.1.2).

Just as they do to the west and sometimes in unison (Fig. 5.45, Section 5.8.7.5) (Dixit and Jones, 1965), "breaks" supervene. Improved weather

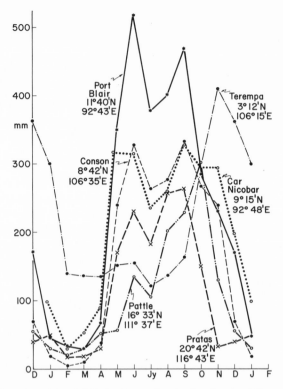

Fig. 5.46. Mean monthly rainfalls for Pratas, Pattle, Con Son, and Terempa in the South China Sea, and for Port Blair and Car Nicobar in the Bay of Bengal.

coincides with a middle-tropospheric ridge. Possible causes of "breaks" seem to be as varied as over India.

Figure 5.38 indicates that rain zones and intervening fair zones more or less follow the sun. However, Fig. 5.46, showing mean monthly rainfalls at four small islands in the South China Sea, and those at two island stations in the Bay of Bengal, reveals no such thing. Although rates of change vary, the curves are in phase; the rainfalls increase to maxima in June and then decrease to secondary minima in July. Mean monthly cloudiness (Sadler, 1969) and daily synoptic and weather satellite charts never show single weather systems extending over the 2300 km separating Terempa from Pratas, let alone into the Bay of Bengal, nor do daily rainfalls at the six stations fluctuate in unison. Apart from the local effects of orographic shadowing (Section 5.8.7.7), India, Burma, and the Indochina peninsula display a well-defined rainfall maximum in July, coinciding with a secondary minimum at the island stations. The continent and adjacent seas are in the same monsoon system—why the difference?

I think the answer lies in the large-scale monsoon circulation (Section 5.8.7.1). The summer monsoon reaches peak strength in July, when the heat trough and the overlying upper-tropospheric ridge are most intense (Fig. 5.3). Since upward motion and rainfall are at a maximum over land, subsidence over the oceans to south and west must also be at a maximum. Over the South China Sea and the Bay of Bengal rains are enhanced as the monsoon strengthens, but a brake—over-ocean subsidence—is also being applied. Even though the brake acts weakly and fitfully so close to the continent, it apparently becomes strong enough on the average in July to produce a secondary rainfall minimum.

(*vi*) *Weather north of the monsoon trough.* Between 30 and 32N average rainfall diminishes after the Mei-Yü ceases (Kao *et al.*, 1962). The monsoon trough does not reach far enough north for disturbances in it to reverse the trend. South of 30N, after a relatively dry second half of July, typhoons and other tropical disturbances cause a secondary maximum in the first half of August.

Amounts dwindle rapidly as the monsoon trough returns south. The first cool, dry winds of approaching autumn reach central China and North Vietnam by late August, a few weeks before the subtropical westerly jet is established south of the Himalayas and along the Yangtze Valley (Section 5.8.1). Only during this "Indian Summer" (Duncan, 1930; Kao, 1958) does average cloudiness over China decrease below 4 oktas.

5.8.7.7   *Interior Plains.* Over peninsular India between the Eastern and Western Ghats (Trewartha, 1961), in central northern Burma, in eastern Thailand, and in eastern Cambodia, rainfall is less in July and August than in early or late summer. The regions are enveloped in moist air from May through October. However, in midsummer, as compared with early or late summer, stronger surface winds enhance subsident drying leeward of the ranges and stronger vertical wind shear inhibits thunderstorm development (Iyer, 1931) in what is predominantly a *showers* regime (Section 4.1.3.2).

5.9   Southern Hemisphere

Limited research on the cognate monsoon regions of southeastern Africa and northern Australia has revealed similarities to the more intensively studied Northern Hemisphere monsoon regions. In an attempt to fill some of the gaps in published information on the Southern Hemisphere regions, I shall therefore analogize.

Topographically, both regions are relatively smooth. In Africa, the Great Central Plateau, rising 1–2 km above a narrow coastal strip, is broken by mountain chains bordering the Great Rift. Madagascar has a rugged mountain spine. Northern Australia possesses very little relief except for the coastal mountains of northeastern Queensland.

5.9.1   APRIL TO OCTOBER

Surface trade winds from an easterly quarter flow out from migrating anticyclones to the south. General sinking motion is maintained by convergent upper-tropospheric westerlies north of the jet stream. Prevailing weather is dry and nearly cloudless except where the trade winds strike mountainous coasts.

Although shallow polar outbreaks may not reach the Great Central Plateau in Africa, orography hinders neither deep polar outbreaks (Frost, 1969) nor large amplitude troughs in the polar westerlies from penetrating both regions. The troughs, which usually move steadily from west, are accompanied by northeastward-moving cold fronts (Fox, 1969; Southern et al., 1970). Occasionally rain is considerable east of a trough line (Fig. 5.47), particularly at the beginning and end of the cool season, and frontal thunderstorms and hailstorms may occur. The troughs are weakest in midwinter when available moisture is least and when the polar anticyclones are farthest north and tend to stagnate over the continents. Over Africa the troughs attain maximum amplitude in April (Thompson, 1965) (Section 5.4.1.1).

A satisfactory explanation is lacking for shifts in the mean latitude of the upper-tropospheric subtropical ridge over southern Africa (Frost, 1969). The ridge lies farther from the equator in July than in either April or October.

The Asian summer monsoon appears to significantly affect the Southern Hemisphere circulation (Section 7.1). East of the central South Indian Ocean, the jet stream accelerates, and a persistent winter-cloud zone extends east-southeastward from New Guinea to Fiji.

5.9.2.   NOVEMBER TO MARCH

In both regions, in the height of summer, east–west troughs lie poleward of the maximum rain belts along about 15–20S. In comparison with the Northern Hemisphere summer, the troughs are neither as dry as the heat trough west of 70E nor as wet as the monsoon trough east of 70E. In the Southern Hemisphere there is no Himalaya–Tibet heat source to fix the pattern of upper-tropospheric divergence (Section 5.1.2) or to restrict or extend the poleward limits of rain (Table 5.11).

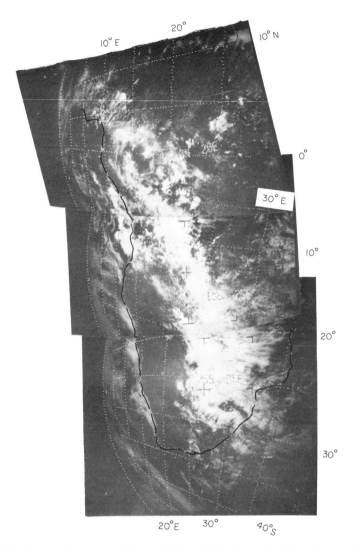

Fig. 5.47. Clouds photographed from ESSA 3, 17 April 1968 (Orbit No. 7071). The cloud zone was moving northeastward ahead of a sharp trough in the polar westerlies. Rain from showers and thunderstorms occasionally exceeded 50 mm in 24 hr (Fox, 1969).

Rain falls in the anticyclonically curving " drift " flow south of the equator (Section 3.10, Fig. 3.33), being especially heavy over northwestern Madagascar where it is reinforced by orography. Even over eastern Madagascar, January rainfall is greater than July rainfall. Although prevailing onshore winds are twice as strong in the latter month, the air is generally rising in

Table 5.11
Poleward limits of summer rain (average latitude of 100 mm monthly isohyet)

North Africa (July)	14N
South and Southeast Asia (July)	33N
Southeastern Africa (January)	22S
Northern Australia (January)	21S

January and sinking in July (Depperman, 1941) (Sections 5.5.2, 5.8.2.1, and 5.8.7.5).

Disturbances resembling monsoon depressions and subtropical cyclones develop (Southern, 1969; Kelly, 1964), with rain concentrated mainly on the low-latitude side of the tracks.

Sensible and condensation heating over southeastern Africa creates an upper-tropospheric high over the continent separated from the Indian Ocean subtropical ridge by a trough along the Mozambique Channel. Thus tropical cyclones approaching Africa from the east almost always recurve before reaching the continent. Rarely do they strike Mozambique, whereas an average of three affect Madagascar each year (Chaussard, 1960) (Fig. 3.6).

Because of condensation heating released by heavy rains over Indonesia, the mean upper-tropospheric ridge is 5° closer to the equator over northern Australia than over southeastern Africa. With presence of the Indonesian chain preventing development north of 10S, tropical cyclones over the Timor and Coral Seas form in higher latitudes and recurve in lower latitudes than is true in other tropical cyclone regions. Consequently, the Gulf of Carpentaria is seldom struck. Perhaps the isohyets between 120 and 125E dip southward toward the desert because of rare cyclones moving directly inland.

The character of summer rain over both regions changes as the season advances. Frequency of thunderstorm days reaches a maximum in midsummer corresponding to Portig's western African type (Section 4.1.1). Many stations in Zambia and Madagascar average more than 20 days in January. However, as over northwestern India (Fig. 4.2), rainfall increases at a greater rate than does thunderstorm frequency. In midsummer, Southern *et al.* (1970) observed periods of *rains* at Darwin with westerlies up to 7 km overlain by strong easterlies, widespread rain having small diurnal variation, below-average thunderstorm activity, a deep moist layer, and relatively stable lapse rate. Similar situations are likely over southeastern Africa.

5.9.3 TRANSITIONS

Though these are ill defined, they usually eventuate during April and May and during October and November. Transformations between near-equatorial troughs and monsoon troughs are usually discontinuous (Thompson, 1965).

Climatological dissipation and new development a thousand or more kilometers distant is typical of the northern Australian region, which in this respect, resembles southern Asia with the Timor and Arafura Seas corresponding to the Arabian Sea and the Bay of Bengal (Sections 5.8.1 and 5.8.4.2). Southeastern Africa, where trough jumps are more in the nature of "noise," resembles northern Africa (Section 5.7).

... there are several Ingenious Instruments invented and improv'd, as Barometers, Hygrometers, Thermometers, etc. by which Men of Ordinary Capacities may pretty easily Prognosticate the Temperature of the Air, and consequently the several Changes and Alterations of Weather.

John Pointer, *A Rational Account of the Weather*, 1738.

6. Synoptic Analysis and Short-Period Forecasting

According to Godske *et al.* (1957), *analysis* "consists in identifying the weather situation with a combination of the tropospheric models . . . in such a way that all, or maximum amount of, the representative observations on the map fit in with the chosen system of models." Consistently accurate *forecasting* cannot be achieved—whether subjective and intuitive or made through numerically iterated solutions of the equations of motion in the largest computers—without accurate analysis. In present-day terminology, the initial state of the atmosphere must be accurately defined as a prerequisite for successful prognosis.

In the remainder of this section, I first discuss analysis and attempts to make it quantitative, and then corresponding efforts at qualitative and quantitative forecasting for periods of about 24 hr.

6.1 Synoptic Analysis

In Chapter 3, I described various synoptic-scale components of the monsoon suitable for use as models in an analysis system. Where the components are well developed, experienced analysts have no difficulty in delineating them and the method of analysis is not important. However, over most of the synoptic chart equatorward of 20°, models cannot be readily identified for most of the year, despite the fact that large weather gradients exist. Then, observational representativeness becomes critical while skillful application of concepts of three-dimensional and time continuity leads to recognition of models which would otherwise have escaped detection. The *method* of analysis now becomes important, and here the lines have been fairly clearly drawn in past years between the proponents of pressure analysis and the proponents of kinematic analysis. In what follows I am concerned only with the relative merits of the two methods when models cannot easily be identified.

6.1.1 PRESSURE ANALYSIS IN THE MONSOON AREA

Pressure analysis is justified largely because the pressure, or force field, depicts a fundamental dynamic property of the atmosphere essential to all attempts at applying the equations of motion to obtain a future configuration of the force field, and, hopefully, a weather forecast. Also, surface pressure can be more accurately measured than any other variable (if a mercurial barometer is used). Isobaric analyses can be quickly performed. The product of a skilled analyst appears superficially similar to that which an unskilled analyst can achieve. This unscientific and tacit justification arises from the fact that forecast offices are usually understaffed and have many under-trained meteorologists.

Nevertheless, let us assume adequate staffing and training. Can pressure analysis provide the basis for an eventual numerical–dynamical forecast system? Presumably it can, but only if the existing pressure field can be quantitatively related to the existing field of three-dimensional motion and, through that, to the weather distribution.

6.1.1.1 *Surface Pressure Field.* Gordon and Taylor (1970) showed that for the pattern of surface divergence (which roughly reflects the distribution of weather) to be reproduced adequately from the sea-level pressure distribution, errors need to be kept below about half a millibar. The coefficient of friction should also be approximately constant. These criteria can probably be satisfied on board ship or on coral islands. Over the continents, errors in determining station heights, local orographic distortions, and uncertainties in reductions to sea level combine to make the goal of sufficiently representative and accurate mean sea-level pressure observations impossible to achieve.

6.1.1.2 *Pressure Field Aloft.* The long-held hypothesis that the geostrophic assumption is seldom valid in low latitudes was supported in a study by Ananthakrishnan and Thiruvengadathan (1967), who compared pressure-height gradients and rawins at standard levels over southern India (Table 6.1). The same type of instrument was used at all the stations. Since curvature accounted for only a small fraction of the deviations from unity in the ratio

Table 6.1

Ratio of median zonal-geostrophic to median zonal-observed winds over southern India[a,b]

Level (mb)	Summer				Winter			
	8.5N–13N		13N–21N		8.5N–13N		13N–21N	
	n	u_g/u	n	u_g/u	n	u_g/u	n	u_g/u
850	333	0.60	290	0.99	195	2.70		
800	328	0.55	283	1.05	203	2.27		
700	323	0.39	236	0.99	207	2.41		
300	286	2.13	277	1.39			178	0.95
250	251	2.04	267	1.41			151	0.98
200	215	1.85	225	1.54			120	1.06

Range at 850 mb:

Summer: 2.70 to -0.72
$\quad\quad\quad$ ($u = 10-15$ m sec^{-1})
$\quad\quad\quad$ 4.55 to -1.25
$\quad\quad\quad$ ($u = 5-10$ m sec^{-1})
Winter: 11.11 to -0.50
$\quad\quad\quad$ ($u = 5-7.5$ m sec^{-1})

[a] From Ananthakrishnan and Thiruvengadathan (1967).
[b] n = number of observations.

u_g/u, one must conclude that there is not much point in estimating winds from the pressure-contour field and vice-versa.

Were both wind and pressure observations sufficiently accurate they could reveal details of the all-important ageostrophic or accelerational components of atmospheric motion, the key to understanding weather and weather changes. Unfortunately, however, over southern India and elsewhere in low latitudes (Ramage, 1964b), the usual errors in pressure-height computations resulting from errors in radiosonde measurements (World Meteorological Organization, 1957; U.S. Committee for GARP, 1969), lead to inaccuracies in estimating the geostrophic wind comparable in magnitude to the wind itself.

One must conclude that most of the time in the monsoon area the pressure field can neither be quantitatively nor qualitatively related to the three-dimensional field of motion nor to the distribution of weather.

" . . . It is clear that only a ruthless analyst who is prepared to ignore any contour height which does not fit in with some preconceived theory, could trace the gradual course of change in horizontal patterns from chart to chart " (Frost, 1969).

6.1.2 KINEMATIC ANALYSIS IN THE MONSOON AREA

In general, surface winds observed at sea and on coral islands and upper winds measured by radar are sufficiently accurate and representative for kinematic analysis to identify models which would be missed in pressure analysis. However, the time and skill required for a good kinematic analysis considerably exceed corresponding requirements for pressure analysis. This is not the place to describe the various kinematical analysis techniques; most tropical meteorological services seriously concerned with analysis, follow the methods developed by Sandström (1909), and Bjerknes *et al.* (1911), and where possible supplement them by pressure analyses. Through judicious use of time and space cross sections, meteorograms, and aerological diagrams (Palmer *et al.*, 1955) they aim at three-dimensional and time continuity.

The diagrams in this book evidence my penchant for kinematic analysis and my conviction that it is *qualitatively* superior to pressure analysis in delineating synoptic models. Nevertheless, even kinematic analysis, close to the equator in the doldrums where quasi-stationary disturbances predominate, often fails to reveal recognizable models.

6.1.3 WEATHER SATELLITES

Much of this book could not have been written prior to 1960 when TIROS I was orbited. The rate at which rapidly developing satellite technology is transforming meteorology suggests that anything I say about weather satellites may be outdated by the time you read this. Conceivably, the novel distance and area-integrating powers of satellites will eventually enable us to overcome the bugbear of unrepresentative and inaccurate point observations and to succeed in quantitatively relating vertical motion to weather, even in the doldrums. For the present though, this goal is beyond reach. All facets of synoptic analysis, especially model identification and tracking, benefit from weather-satellite data.

From the mass of references I have space to mention only a selected few. A good general survey (National Aeronautics and Space Administration, 1966) introduced techniques for interpreting the data from earth-orbiting satellites and analyzing them in combination with earthbound data (Sadler, 1962; Hanson, 1963; Widger *et al.*, 1964; Anderson *et al.*, 1966). Pyle (1965), Booth and Taylor (1969), and Leese *et al.* (1969) described the data and how they are made available to researchers. Inexpensive ground stations (Vermillion, 1969) can receive clear detailed pictures direct from satellites equipped with Automatic Picture Transmission (APT).

The new "earth stationary" synchronous satellites (McQuain, 1967;

Warnecke and Sunderlin, 1968) make it possible to take pictures of the same one-quarter of the earth's surface at intervals as small as 20 min. Now diurnal changes can be studied, and cloud motions are being used to deduce the wind and circulation patterns (Fujita *et al.*, 1969).

Satellites will be incorporated into the successful GHOST system (Lally *et al.*, 1966) to track superpressure balloons along constant pressure levels. Vertical temperature profiles in the atmosphere have been calculated from measurements made by an infrared spectrometer on the Nimbus III satellite (Wark and Hilleary, 1969). A Michelson interferometer spectrometer carried on the same satellite has provided data from which vertical profiles of temperature, water vapor, and ozone have been derived (Hanel and Conrath, 1969).

6.1.4 Objective Analysis

Most meteorologists subscribe to the tenet that everything objective an experienced analyst can do, a sufficiently large computer can be programmed to do. In the tropics little is gained from using pressure gradients to estimate winds (Table 6.1). Consequently, Bedient and Vederman (1964) developed a program to analyze the wind field directly from wind measurements. Analysis, which incorporates an error-detection test, starts with a "first guess"—the analysis for the preceding synoptic hour (modified by applying a trend toward climatology), expressed as winds at intersections of a grid with a mesh length at the equator of 5° of longitude. These winds are then "corrected" by comparison with observed winds. The vector correction, $\mathbf{K} = -$(weighting factor) $(\mathbf{V}_i - \mathbf{V})$ where the weighting factor is defined as

$$[(D/2)^2 - d^2]\big/[(D/2)^2 + d^2] \tag{6.1}$$

Here d is the distance between the grid point and the observation point and $D/2$ is the distance at which the weight is zero. $\mathbf{V}_i$ is the "first-guess" wind at a grid point.*

In one method the computer scans a circle of radius $D/2 = 3$ grid lengths (Carlstead, 1967), centered on the grid point. Differences between $\mathbf{V}_i$ and wind observations within the circle are individually weighted according to Eq. (6.1). The average of the vector corrections so derived is applied to $\mathbf{V}_i$ to give a new $\mathbf{V}_i$ or a "second-guess" wind. The procedure is repeated three times, with $D/2$ being successively diminished to 2, 1, and 0.5 grid lengths.

---

* The computer is programmed to operate on the u and v components (scalars) of the winds.

The final V_i is the new analyzed wind at the grid point. According to Bedient and Vederman, (1964) " . . . the iterative process . . . has the effect of adjusting the first approximation for the large-scale features of the circulation and later approximations for smaller and smaller features on successive scans."

A recent improvement (Carlstead, 1969) makes the scan radius a function of the density of data surrounding a grid point. Smoother analysis results.

The analysis, contained in the computer in the form of winds at grid points, can be readily incorporated in persistence-climatology wind forecasts (Section 6.2.2.2) and the results printed out in forms most useful to flight operations.

People still like to *look* at weather maps; they are readily humored by programming the computer to direct a graphical-incremental plotter to draw vectors at each grid point and even to trace streamlines (Davis, 1969) (Fig. 6.1).

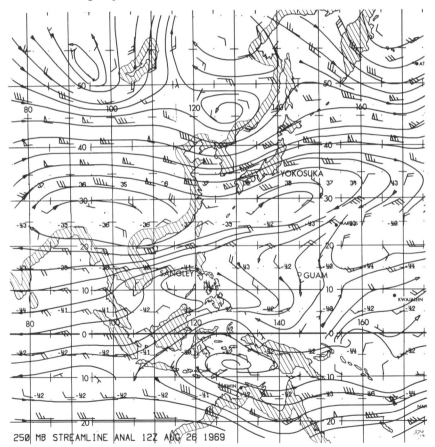

Fig. 6.1. Sample 250-mb wind analysis, with machine plotted winds and streamlines, made at Fleet Weather Central, Pearl Harbor, Hawaii (Davis, 1969).

The analysis can be kept up to date. Aircraft wind observations, often not at synoptic times, are punched on cards and inserted in the computer which takes account of them and immediately adjusts the analysis.

The computer must be supervised. A monitoring meteorologist checks whether erroneous data slip by automatic screening procedures. He may also modify the computer analysis in the vicinity of tropical storms (Rush, 1967) or in the light of nondigital data such as satellite pictures.

On the assumption that vertical wind shear is a relatively conservative quantity, it is used in the computer to screen new data for possible errors. Nevertheless the system does not yet satisfactorily incorporate scanning in the vertical and in time: second nature to the good analyst and key to high quality work.

6.1.5 Quantitative Applications of Kinematic Analysis

It is a truism that vertical motion is the key to weather. It is also true that vertical motion has never been measured on a synoptic scale. In low latitudes where local density changes can be ignored, the equation of continuity applied to isobaric surfaces is given by

$$\partial \omega / \partial p = -\nabla \cdot \mathbf{V} \tag{6.2}$$

Integration with respect to pressure leads to

$$\omega_i - \omega_0 = -\int_{p_0}^{p_i} \nabla \cdot \mathbf{V} \, dp \tag{6.3}$$

where subscript 0 refers to the earth's surface. ω_0 is either zero if the surface is flat or can be determined from wind speed and orography if the surface is rough. The average divergence in the layer between p_0 and p_i then leads to a determination of ω_i at the top of the layer. The same procedure is followed for the next layer with ω_i replacing ω_0 in the equation, and so on.

Although some air moves through the tropopause, it is nevertheless a layer of minimum vertical motion, particularly in low latitudes. Therefore since vertical motions are damped out near it, ω_{100} (Fig. 6.2) should be approximately zero.

The surface-pressure tendency very rarely exceeds 3 mb day^{-1} in the monsoon area. Reasonably vigorous divergence of 10^{-5} sec^{-1} acting on the layer between 1000 and 700 mb would produce a pressure fall of 10 mb hr^{-1} at the surface. Therefore compensating convergence must occur at higher levels in order to keep the surface pressure change in the two-orders-of-magnitude-lower range generally observed.

The criteria that the tropopause approximates a material boundary to synoptic-scale vertical motions and that net mass divergence in the tropospheric column approximates zero can be applied to divergence and vertical motion computations to evaluate their validity.

With *accurate* and *representative* wind soundings spaced at distances which are small compared to the size of synoptic systems, an analyst could apply Eq. (6.3) and determine the distribution of vertical motion. Then if he knew the distribution of moisture from radiosoundings he might hope to relate vertical motion quantitatively to cloud amount and depth and even to rainfall.

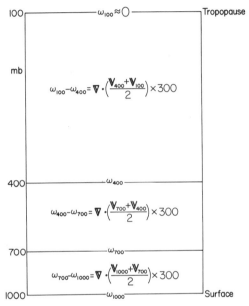

Fig. 6.2. Approximate relationship between divergence and vertical motion (ω) in a three-layer troposphere (1000–700 mb; 700–400 mb; 400–100 mb).

6.1.5.1 *Surface Divergence.* How representative of the surrounding area circulation is a wind observation? Surely a wind measured automatically with good equipment at a well-exposed site on a tiny coral island is representative of the wind over the surrounding ocean. During the Line Islands Experiment (Zipser and Taylor, 1968) three observing stations were established on Palmyra Island, comprising 6 km² of land 3 m above sea level at its highest point, and 23 km² of lagoon.

All the stations, at similarly open, well-exposed sites, were identically equipped with hyetographs and anemographs. The Causeway station was situated 1.4 km east and the Barren Island station 4.3 km eastsoutheast of the Airfield station.

Table 6.2 compares the frequencies of measured rainfall and the average wind speeds in the hour preceding each of the four synoptic observing times: 0000, 0600, 1200, and 1800 GMT, for the period 17 March through 18 April 1967. Considering the Airfield and Barren Island stations, on 7 occasions rain occurred at the former and not at the latter, on 2 occasions at the latter and not at the former and on 15 occasions during the same hour at both. Wind averaged 50% stronger at Barren Island than at Airfield. When this unrepresentativeness component is added to an expected measurement error of ± 3 m sec^{-1} or even a hoped for measurement error of ± 1 m sec^{-1} (U.S. Committee for GARP, 1969), then the fact that synoptic-scale divergence seldom surpasses 10^{-5} sec^{-1} in the tropics means that on most occasions not even the sign of the surface divergence can be determined.

6.1.5.2 *Vertical Motion.* If surface divergence cannot be accurately computed then it is unlikely that vertical motion can be accurately computed from calculations of divergence in atmospheric layers. Rex (1958) reported an attempt based on careful kinematic analyses of high-quality rawin observations made at stations about 1000 km apart in the central Pacific. Although

Table 6.2

Rain frequency and wind speed measured at three stations on Palmyra Island (5°53′N, 162°05′W) from 17 March through 18 April, 1967

	Occasions of rain measured in preceding hour	Average wind speed (m sec^{-1}) in preceding hour
0000 GMT		
Airfield	5	4.4
Causeway	3	5.5
Barren Island	3	5.9
0600 GMT		
Airfield	6	3.8
Causeway	6	5.3
Barren Island	4	5.9
1200 GMT		
Airfield	5	3.9
Causeway	2	5.5
Barren Island	6	6.3
1800 GMT		
Airfield	7	4.1
Causeway	7	5.6
Barren Island	4	6.4

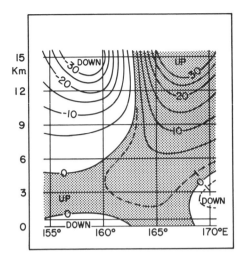

Fig. 6.3. Vertical motion (cm sec^{-1}) is along 11 N at 0300 GMT, 22 June 1955. The region of upward motion is stippled (Rex, 1958).

cloudy regions roughly coincided with computed upward motion, Fig. 6.3 reveals very serious errors—strong vertical motion at the tropopause. Even greater errors occur in the monsoon area (Saha, 1968), especially over rugged continents with large horizontal heating and friction gradients.

Thus instrumental errors and lack of representativeness prevent any quantitative relationship being established between divergence-derived vertical motion and weather, while qualitative relationships are evident only in vigorous weather systems with large wind gradients. Increased measurement accuracy would improve matters. But in the light of the Palmyra experience, smaller-meshed station networks would exacerbate the difficulty by proportionately increasing the contribution to "noise" of instrument shortcomings and unrepresentativeness.

Much of the "noise" inherent in synoptic wind data is eliminated by averaging; by this means circulation and vertical motion have been computed for "average" typhoons and subtropical cyclones and related meaningfully to corresponding "average" cloudiness and rainfall (Sections 3.3 and 3.5).

6.1.5.3 *Vorticity and Weather.* Beyond the equatorial region rather weak vorticity centers, which persist from one synoptic chart to the next, can sometimes be associated with correspondingly persistent weather patterns. The reason is not hard to find—in a cyclonic circulation friction causes convergence toward lower pressure; in an anticyclonic circulation friction causes divergence away from higher pressure. A surface low overlain by an upper-tropospheric high is likely to be found in a bad-weather region, for continuity demands that upward motion must occur between the two.

Summing up, in monsoon regions of small wind or pressure gradients no known method of synoptic analysis or combination of methods can consistently account for the weather distribution. Hindsight often miraculously reveals unsuspected models which then unaccountably dissolve under the acid test of operations.

6.2 Short-Period Forecasting

In the monsoon area, two methods of short-period (up to 24 hr) weather forecasting are used, separately and in combination—forecasting based on careful synoptic analyses and forecasting based on probabilities determined through statistical analyses. The skilled, successful forecaster critically appreciates the merits and drawbacks of both.

6.2.1 SHORT-PERIOD FORECASTING BASED ON SYNOPTIC ANALYSIS

Synoptic analysis identifies and follows well-defined circulation models from chart to chart and along adjacent pressure levels. The more careful and skilled the analyst is, the smaller will be the areas on his charts lacking identifiable models. Where weather distribution cannot be accounted for by analysis, analysis cannot aid forecasting. The question is, how to forecast what will happen to the painstakingly identified models? Many meteorologists have tried to turn the vorticity equation to this purpose.

6.2.1.1 *The Vorticity Equation.* Neglecting the effects on vorticity due to friction, baroclinity and tilting, the rate of change of absolute vorticity, following the motion can be written

$$d\zeta_a/dt = -\zeta_a \nabla \cdot \mathbf{V} \tag{6.4}$$

Determining the relative vorticity $\zeta = \partial v/\partial x - \partial u/\partial y$ is not as difficult as determining the divergence for in the former $\partial v/\partial x$ usually has the same sign as $-\partial u/\partial y$ whereas in the latter $\partial u/\partial x$ and $\partial v/\partial y$ are about the same size but usually have different signs. Thus not only the sign, but also the approximate magnitude of the vorticity can be computed from any reasonably dense network of wind-measuring stations. Tropical meteorologists have long been intrigued by the possibility of using the vorticity equation to circumvent the insuperable problem of measuring the divergence (Riehl, 1954) to lead directly to a *forecast* of divergence and through that, of weather.

At first sight the prospects appear good. One observes on following air-parcel trajectories through a sequence of synoptic charts that in the lower troposphere, parcels moving through bad weather areas gain absolute vorticity (increasing convergence), while above them in the upper troposphere moving parcels lose absolute vorticity (increasing divergence). Continuity then prescribes the expected upward motion.

If only $d\zeta_a/dt$ could be forecast, all would be well. Providing that a synoptic circulation of unchanging intensity travels at a markedly different speed from that of the air associated with it, as is usual with long-wave troughs in the polar westerlies, easterly waves (Riehl, 1954), and typhoons, then the sign of $d\zeta_a/dt$ and consequently of $\mathbf{V} \cdot \mathbf{V}$ can be readily determined. Fine weather close ahead and to the right of a slowly moving typhoon can be thus accounted for. In surface air overtaking the typhoon $d\zeta_a/dt > 0$ whereas on leaving, $d\zeta_a/dt < 0$.

In practice however, the problem is far from simple, for relative motions of air and circulation systems often change unexpectedly as do the intensities of the systems, while Hastenrath (1968) demonstrated that, near the equator, the frictional and tilting terms in the vorticity equation cannot be neglected if circulations are to be adequately described.

Buajitti (1964) statistically analyzed 24-hr vorticity changes over Thailand for five years' of summer monsoons. Upper winds measured at 11 stations provided the data. At the 1.5-km level he found (Table 6.3) that where a

Table 6.3
Twenty-four hour changes in relative vorticity (ζ) and weather in the subsequent 24 hr over Thailand[a]

ζ Change	Weather
$>8 \times 10^{-5} \text{ sec}^{-1}$	Rain
$2\text{–}6 \times 10^{-5} \text{ sec}^{-1}$	50–60% rain
$<-8 \times 10^{-5} \text{ sec}^{-1}$	No rain

[a] From Buajitti (1964) for triangular areas of about 25,000–30,000 km^2.

large positive change in vorticity occurred, rain was measured in the subsequent 24-hr period; where a large negative change occurred, no rain fell in the subsequent 24-hr period. His results are physically reasonable; summer circulations over Thailand are usually quasistationary and so divergence of air moving into a system whose vorticity had recently significantly changed would be expected to change correspondingly but with some lag. Buajitti reported that large vorticity changes occurred rather rarely, but he did not indicate whether they could have been readily recognized in the course of careful analysis.

(i) Barotropic forecast model. The instantaneous change in vorticity following the motion

$$d\zeta_a/dt = \partial\zeta/\partial t + \mathbf{V}\cdot\nabla\zeta_a + \omega\,\partial\zeta/\partial p \tag{6.5}$$

if averaged through the atmospheric column or if referred to a level of non-divergence (usually between 500 and 600 mb) approximates zero, while $\omega\,\partial\zeta/\partial p$ is very small. The vorticity equation then becomes

$$\partial\zeta/\partial t + \mathbf{V}\cdot\nabla\zeta_a = 0 \tag{6.6}$$

which expresses the conservation of absolute vorticity.

In middle-latitude barotropic forecasts $\mathbf{V}\approx\mathbf{V}_g$ and*

$$\zeta\approx\zeta_g = (980/f)(\nabla_p^2 h)$$

In the tropics, however, $\mathbf{V}_g \not\approx \mathbf{V}$ (Table 6.1). To circumvent this obstacle, Vederman et al. (1966) experimented with replacing the pressure–height field with the stream-function field which describes the nondivergent part of the wind.

Assuming that $\mathbf{V}_\psi\approx\mathbf{V}$, Eq. (6.6) then becomes

$$(\partial/\partial t)\nabla^2\psi + \mathbf{V}_\psi\cdot\nabla\zeta_a = 0$$

or

$$(\partial/\partial t)\nabla^2\psi + J(\psi, \zeta_a) = 0 \tag{6.7}$$

In the experiment which was conducted over the central Pacific, ψ was presented on a grid with a mesh size of 556.5 km at the equator and numerical computation in successive 1-hr time steps gave a 24-hr forecast of the distribution of ψ for the 700-, 500-, 300-, and 200-mb levels. Errors in the forecast stream-function winds exceeded errors in routine forecasts. In a later experiment (Vanderman and Collins, 1967) embracing the entire tropical belt, 36-hr forecasts improved slightly on persistence at 300 mb and were slightly inferior to persistence at 700 mb.

In the barotropic model potential and internal energy cannot be converted into kinetic energy and cyclones or anticyclones cannot develop except apparently when absolute vorticity from several sources is concentrated at a point. Besides, boundary errors seriously reduce the accuracy with which

---

* Equation following from Panofsky (1957).

ψ can be determined (Bedient and Vederman, 1964), although extending the boundaries into middle latitudes (Bedient *et al.*, 1967) has partially solved this problem.

The model is even less likely to perform well in the monsoon area where both surface roughness and baroclinity are greater than over the oceanic tropics.

In short, the vorticity equation does help explain circulation–weather relationships in well-defined models and to that extent contributes to good analysis, but its contribution to day-to-day forecasting in the monsoon area is questionable.

It is probably too soon to decide whether the promising diagnostic model due to Murakami (1969) (Section 5.8.2.3) can lead to improved daily numerical weather forecasts in low latitudes.

6.2.1.2   *Extrapolation.* In typhoon- or monsoon-depression track charts, individual storms generally travel along rather smooth curves and at speeds which vary only slowly. Generally, too, the storms only slowly change intensity. Not surprisingly, the most successful monsoon-area weather forecasts are made in the vicinity of such vigorous systems.* The key is extrapolation: identifying and following a trend, while at the same time attempting to explain the trend by reference to analyses.

Detailed pictures of clouds obtained from weather satellites through Automatic Picture Transmissions (APT) (Section 6.1.3) have opened up prospects for greatly expanded use of the extrapolation principle. Often an alert analyst can first detect a trend in these pictures before any evidence is apparent from regular synoptic reports, and by extrapolating that trend (which may be a combination of space and time trends) make an excellent forecast.

Between 12 and 13 August 1964 (Figs. 3.27 and 3.28), for example, the trend toward a weakening subtropical cyclone and diminishing rain was clearly shown by the change in cloud characteristics photographed by the weather satellite. Early recognition of a trend and its extrapolation may be denigrated by the purist as nothing more than "persistence forecasting" but it demands great skill and efficient acquisition of numerous data.

At least during the summer monsoon, the forecaster has a better chance of anticipating improving weather than deteriorating weather (Sadler *et al.*, 1968). Perhaps then his attention should be refocussed and his forecasts might safely be more specific during a fine spell and more vague during a wet spell.

---

* However, because of large wind and weather gradients small errors may be less tolerable than large errors incurred in forecasting less-intense systems.

Thus far I have concentrated on analysis as an aid to *weather* forecasting. Wind forecasting is just as important, but since wind and weather forecasts derived from analysis are inseparable, success in one implies success in the other. However, in the next section wind forecasting can properly be treated separately.

A year after I first came to the monsoon area I evaluated local forecasts made by my colleagues and me and found to everyone's consternation and near disbelief, including mine, that our performance was significantly worse than the climatology and that among us experience counted for naught. This startling but not uncommon denouement stemmed from a human reluctance to leave well alone and an urge to forecast change, and from a lack of appreciation for statistics as an aid to forecasting.

Over the past twenty years, statistics have become fashionable and "objective" techniques are now available for making short-range forecasts of almost anything. A judicious blend of statistics and careful synoptic analysis can significantly improve the quality of short-range forecasting in the monsoon area, for the former is most effective in just those very amorphous situations which defy the model-fitting methods of the latter.

6.2.2   SHORT-PERIOD FORECASTING BASED ON STATISTICS

At Khartoum in winter or at Cherrapunji in summer, climate so closely resembles day-to-day weather that the long-term mean would almost always be a successful short-period forecast. However, in most parts of the monsoon area, weather is too changeable for such a simple scheme to be effective.

6.2.2.1   *Local Climatology.* Forecasts for restricted areas such as cities or airfields can be greatly aided by means or frequency analyses of meteorological variables. Information on the chances of a particular meteorological event—a thunderstorm, for example—occurring in a particular month can be used intelligently to modulate any trend detected from synoptic analysis. Considering diurnal variation curves further sharpens forecast accuracy. Such statistics are most effective near the equator where moving weather systems are rare, and over land with rugged terrain, where orography exerts a strong effect (Section 4.1.3.1, Fig. 4.6).

6.2.2.2   *Combination of Climatology and "Persistence."* In many parts of the monsoon area tolerable results are usually obtained when tomorrow's weather is forecast to be the same as today's.

In middle latitudes meteorologists found that short-period *upper-wind forecasts* combining climatology and persistence improved on mere climatology or persistence and compared favorably with wind forecasts based on synoptic analysis and dynamic reasoning (Durst, 1954; Durst and Johnson, 1959). In simplified form the equation for prediction of wind at a point is

forecast vector wind $= (1 - r_t)$ (climatological resultant vector wind)

$$+ r_t \text{ (vector wind observed at forecast time)} \quad (6.8)$$

where r_t, determined from past data, is the correlation coefficient between vector winds and vector winds lagging them by the period of the statistical forecast. Over the tropical Pacific good 24-hr upper-wind forecasts were first achieved by making $r_t = 0.5$ (Lavoie and Wiederanders, 1960); the wind was forecast to be the vector mean of the mean resultant wind and the observed wind.

Later, when forecasts were calculated by computer, r_t was determined separately for each 5° latitude–longitude grid point. These forecasts are the best available for low latitudes while the hemispheric numerical forecasts made by the National Meteorological Center, Washington, D.C., are the best available for middle latitudes. The two can profitably be "merged" between 20 and 30N.

After comparing performances by several statistical and other methods, Chin and Leong (1964) concluded that since simple combination of half-persistence with half-climatology gave 24-hr upper-wind forecasts at Hong Kong within 0.5 m sec^{-1} of the best of the other forecasts, it could scarcely be improved upon.

Combining persistence and climatology in making forecasts need not be limited to the upper-wind field. McCabe (1961) showed that subjective terminal forecasts of cloud ceiling, visibility, and precipitation could be thereby improved and also that the concept could be successfully applied to the problem of forecasting typhoon movement.

Of course the very nature of the method precludes successful forecasts of extremes or sudden changes.

6.2.2.3 *Multiple-Predictor Statistics.* Synoptic experience leads meteorologists to associate events with corresponding earlier events and to attempt physical explanations for the association. For example, surface cyclogenesis over the East China Sea is usually followed by a cold surge across southern China. Also, a cold surge often follows a surface-pressure rise over southern Siberia. Cyclogenesis and anticyclogenesis would appear to "predict" a cold outbreak.

Two statistical methods are now widely used to determine whether such observed associations are statistically significant and to combine those that prove to be, into objective forecast methods. To reduce computer time, experienced analysts first identify meteorological elements which appear to vary directly or inversely as the element to be predicted subsequently varies. When a sufficient number of individual cases have been assembled, each "predictor" is linearly correlated with the "predictand." If more than one significant correlation coefficient results, then either a forecast equation is obtained through a stepwise regression (screening) procedure or a least-squares procedure leads to a graphical forecast.

(i) *Stepwise regression (screening)*. The procedure, applied to weather prediction by Miller (1958), was described by Aubert *et al.* (1959). First, the potential predictor most highly correlated with the predictand is identified. Through orthogonalizing, the predictand variance accounted for by this predictor is removed. In the next step a second predictor most highly correlated with the residue of predictand variance is identified, and the variance accounted for by it is removed. Third and subsequent steps follow, stopping before the step which would reduce the variance by less than a predetermined amount. The forecast equation (compensated for differences among predictor units) comprises an origin adjustment, followed by the predictors, each multiplied by its correlation coefficient. As an example (Arakawa, 1963), a prediction equation for the longitude of a South China Sea typhoon center 12 hr hence is

$$\lambda_{+12} = 97.9 + 1.0122\lambda_0 - 0.1562x_4 + 0.1717x_{17} - 0.1151x_6$$

where λ_0 is the present longitude of center, and $x_1, x_2, \ldots, x_{25}$ are present surface pressures at the intersections of a 5° latitude–longitude grid centered at the typhoon center (x_{13}). The percentage reduction of variance amounts to 95.3.

Physical connections are usually apparent between the first two predictors selected by screening, and the predictand. However, successive reduction of the variance precludes all but chance physical connections between succeeding predictors and the predictand.

The screening procedure usually needs a computer. However, similar but less rigorous methods have been devised which are not computer dependent. Tse (1966) plotted 700-mb pressure-height gradients against typhoon movement on scatter diagrams. He then introduced the overall synoptic pattern at 700 mb as an additional criterion to subdivide the scatter distributions into a series of regression lines. Although the method involved subjective classification of synoptic situations, it performed as well on dependent data as did equations derived from screening procedures, perhaps because the statistics could be readily explained by physical reasoning.

(*ii*) *Forecasting from statistically determined graphs.* The method, described by Freeman (1961), resembles screening and requires a computer. In successive steps:

1. Curvilinear correlation coefficients are computed between potential predictors and the predictand. A polynomial (usually of fifth power or less) is fitted to the data by the method of least squares, giving the correlation coefficient

$$r^2 = 1 - (\text{root mean square error/standard deviation of the predictand})^2 \qquad (6.9)$$

2. Of the most promising predictors (determined in step 1) the best dual combination is found. Now a curved *surface* is fitted to the data by the method of least squares. In the polynomial of the form

$$Z = a + bx_1 + cy_1 + dx_1^2 + ex_1y_1 + fy_1^2 + gx_1^3 + \cdots,$$

Z is the predictand and x_1 and y_1 the two predictors which together give the least root-mean-square forecast error. The curved surface is then represented by a plotted set of curves which enable the first approximate forecast to be made (Fig. 6.4, bottom left).

3. The procedure is repeated. Z, as determined in step 2, becomes x_2 in a new polynomial. y_2 is the predictor which, when combined with x_2, gives the least root-mean-square forecast error of all the possible new predictors in the store of potential predictors established by step 1. A new set of curves representing the second surface (Fig. 6.4, bottom right) leads to a closer approximation to a correct forecast.

4. Subsequent steps repeat step 3 and bring a correct forecast closer. However, each step reduces the variance less than the preceding step. Therefore the sequence is halted just before the step which would reduce the variance by less than a small predetermined amount (Fig. 6.4, top). Further steps would add complications but not improve forecast performance.

Almost invariably in the tropics, of all possible predictors, the present value of the variable to be predicted is most highly correlated with the predictand. This objectively confirms what shrewd meteorologists have long known, that "persistence" is hard to improve on, whether upper tropospheric winds, typhoon movement, or temperature is being forecast.

6.2.3 SUMMARY

For those who yearn for precision, forecasting in the monsoon area is disappointingly inexact. Nevertheless, a meteorologist, after intelligently and

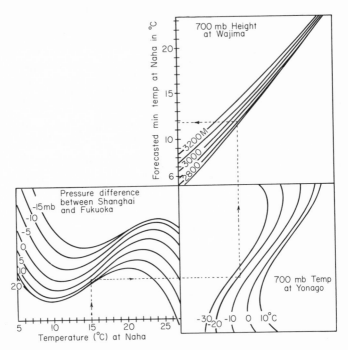

Fig. 6.4. Minimum temperature prediction diagram for Naha (26°12′N, 127°39′E). The dotted line indicates the sequence followed in making a forecast (Nakamura, 1967).

deliberately studying a detailed sophisticated climatology and a sequence of carefully analyzed synoptic and auxiliary charts, can forecast consistently better than chance. A statistical prediction should always be available to him. He should modify that prediction only when he discovers a significant change trend in the charts. When in doubt, stay with statistics.

In a few years, the millennium envisioned by global theoreticians (U.S. Committee for GARP, 1969) may come to pass and changes in the large-scale tropical circulations be quite accurately forecast several days ahead.

Satellites and radar have revealed mesoscale weather gradients, within recognizable circulations, such as typhoons or monsoon depressions, as great as those between the circulations and their environments. Besides, less than 20% of the area of the most intense cyclones is covered by vigorously precipitating clouds. Conversely, where fine weather predominates, near the heat trough, a thunderstorm or dust storm may suddenly surprise a locality.

Thus, successful large-scale forecasting would force monsoon-area meteorologists to sharpen their synoptic and statistical weapons for a concerted attack on the perplexing problems of mesoscale forecasting.

A correlation coefficient is a very sensitive plant, it is much easier to kill one than to make one.

Sir Napier Shaw, *Manual of Meteorology*, 1936.

7. Monsoons and the Atmospheric Circulation; Floods, Droughts, and Trends

In Chapter 2, I mentioned briefly important monsoon-area weather cycles, each comprising a succession of synoptic events, akin to Grosswetter-lagen (Baur, 1951) (Fig. 2.8). With periods of weeks, the cycles embody the major weather anomalies of vital concern to agricultural planning and practice (Blanc, 1965)—floods, droughts, and interannual variations in monsoon rain. Since the start of organized meteorology in the monsoon area, scientists have tried to explain these larger fluctuations and, prodded by the community, to forecast them (Normand, 1953).

Their extents and durations suggest that the cycles (of which the Indian summer-monsoon rain-break cycle (Section 5.8.7.5) is an example) could arise only from interactions between air over the monsoon area and the atmospheric circulation as a whole. Although a start has been made with the Indian summer monsoon, definitive studies of significant seasonal or part-seasonal anomalies have never been completed. But the problems are sufficiently important and eventually tackling them successfully through numerical techniques is sufficiently promising to justify the fragmentary, illustrative discussion attempted in this chapter.

In the remaining sections I discuss:

(1) two monsoon systems: the South Asian summer monsoon and the East Asian winter monsoon, and their possible relationships to circulations across the equator and in higher latitudes;
(2) the great variability of July rain in central China; and
(3) long-range forecasting performance and potential.

7.1 Transequatorial Effects of the South Asian Summer Monsoon

7.1.1 SOUTHERN HEMISPHERE CIRCULATION

Over the Southern Hemisphere the zonal component of the upper-tropospheric winter circulation is far from uniform. From June through August an exceptionally large vertical wind shear extends along 23–25S from the eastern Indian Ocean across Australia and the western half of the South Pacific (Van Loon and Taljaard, 1958). In the resulting jet stream, average speeds exceed 70 m sec^{-1} over Australia (Muffatti, 1964). Over southern Africa, however, the jet stream is ill defined and average winds barely reach 35 m sec^{-1} (Hofmeyr, 1961). Besides, troughs and ridges in the polar westerlies appear in unusual locations. According to Rubin (1955), in winter, Australia should be the seat of a 500-mb dynamic ridge, but instead, a 500-mb trough usually lies along about 120E.

7.1.2 TRANSEQUATORIAL INTERACTION

"... it appears as though the monsoon effect of the Himalayas also extends into the *upper* flow pattern of the southern hemisphere. That the southeast trades of the southern hemisphere over the Indian Ocean merge into the southwest monsoon of summer and of the northern hemisphere is already a well-known fact. A coupling of the circulations of both hemispheres, thus, seems to be indicated " (Reiter, 1963).

In equatorial regions, lower-tropospheric northward flow and upper-tropospheric southward flow over the western Indian Ocean are unmatched for strength and persistence. Trajectories usually closely approximate streamlines. According to the arguments presented in Section 3.10, maximum convergence should occur in anticyclonic flow around the western and southwestern segments of the South Indian Ocean upper-tropospheric subtropical ridge. Resulting subsidence warms the air which then flows alongside the polar westerlies, increasing the meridional temperature gradient (Van Loon and Taljaard, 1958), and generating the jet stream. Thus, in opposite seasons, Himalaya–Tibet indirectly causes the strongest Southern Hemisphere jet stream as well as the strongest Northern Hemisphere jet stream (Section 5.1).

The most persistent winter cloud zone in the Southern Hemisphere tropics (except for stratus sheets where cold water upwells) extends east–southeastward from New Guinea to Fiji (Sadler, 1969). Why should the cloud, predominantly altostratus or nimbostratus (Hill, 1964), tend to persist here? No good local reason coming to mind, it may stem from quasi-stationary dynamic instability in the abnormally strong jet stream to the south (Ramage, 1970).

To extrapolate from climatology, perhaps Grosswetterlagen in the north are linked with Grosswetterlagen in the south. The heavier the monsoon rain over India, the stronger might be the transequatorial circulation in the vertical plane* and the stronger the Australian jet stream. Averages for the same calender month in different years are quite inconsistent with this model, and Grosswetterlagen over India are frequently out of phase with those over Australia.

This disappointing conclusion echoes an earlier failure. In exhaustive studies of the great western Indian drought of 1899, Dallas (1900, 1902) and Eliot (1905) suggested causes of the lack of rain in July which patently could not account for the continued drought in August and September.

We are led inescapably to the unsatisfactory conclusion reached in discussion of "breaks" in the summer monsoon *rains* of India (Section 5.8.7.5), i.e., with zonal and transequatorial circulation modes operating, a particular anomaly in one region might arise from or be accompanied by several different anomaly combinations in other regions. *Climatologically*, the role of Himalaya–Tibet is clear; on the scale of Grosswetterlagen, it merely modulates atmospheric signals from many sources.

7.2   The East Asian Winter and Australian Summer Monsoons

Section 5.8.2.3 described near-simultaneous fluctuations in the vigorous winter Hadley cell extending from central China to Indonesia. Figure 5.20 shows that North Borneo rainfall varied in unison with the circulation intensity. Does a relationship suggested by synoptic sequences obtain on the Grosswetterlage scale?

Meteorological variables are never routinely averaged over "natural" weather periods. Hence to test this hypothesis conveniently requires that at least two individual winter months with sharply differing weather regimes be identified and compared.

7.2.1   JANUARY 1963 AND JANUARY 1964 OVER SOUTHEAST ASIA AND THE WESTERN PACIFIC†

January 1963 and January 1964 were well suited for a comparative study. Over the Northern Hemisphere in the regime of the upper westerlies, *January*

---

* The lower-tropospheric winds at Port Blair (11°40′N, 92°43′E) and the upper-tropospheric winds at Gan (0°41′S, 73°09′E) together provide a rough index of this (Section 5.8.7.5).

† Ramage (1968b).

1963 (O'Connor, 1963) "was memorable for the extreme severity of the cold weather which simultaneously gripped North America, Europe, and the Far East." Over the western and central Pacific the jet stream was much stronger and farther south than normal while an intense blocking ridge persisted over the eastern Pacific. At the surface, both the Siberian high and Aleutian low were unusually intense (Japan Meteorological Agency, 1963). Rains were above normal over Indonesia and the Carolines (the "maritime continent"), being very heavy over eastern Malaysia, and below normal over New Guinea. *January 1964* (Andrews, 1964; Japan Meteorological Agency, 1964) was a much milder month. Over the Eastern Hemisphere

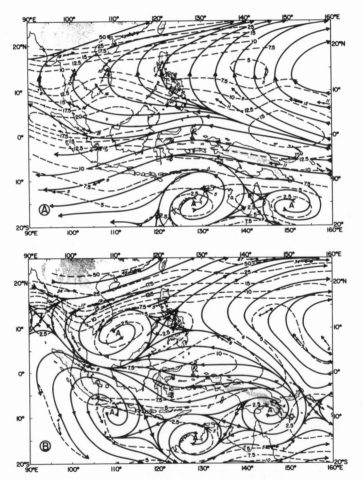

Fig. 7.1. Mean resultant circulations (m sec^{-1}) at 200 mb. (A) January 1963; (B) January 1964.

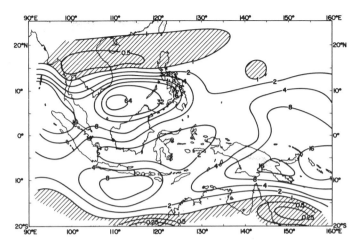

Fig. 7.2. Ratio of January 1963 kinetic energy to January 1964 kinetic energy at 200 mb.

meridional flow prevailed with the subtropical jet in about its normal position. The Aleutian low lay farther east than usual and pressure gradients between it and a near normal Siberian high were only half as large as in January 1963. Rains over the maritime continent were much below normal, except for New Guinea; central Indonesia suffered the worst drought in living memory.

Figure 7.1 depicts mean resultant circulations at 200 mb for January 1963 and January 1964. The circulation in 1963 was much more intense (Fig. 7.2). The poleward component across the subtropical ridge averaged 7 m sec^{-1} in 1963 and only 3 m sec^{-1} in 1964. Lower levels present the same picture (see above and Table 7.1). Over the maritime continent rainfall in 1963 exceeded the 1964 amount by at least 200 mm (Fig. 7.3). If converted to mechanical energy, this difference could account for the difference in circulation intensity between the two months.

Table 7.1

Mean northerly component of the surface wind (m sec^{-1}) over the South China Sea

	January 1963	January 1964
Pratas 20°42′N, 116°43′E	9.9	4.3
Paracels (Sisha) 16°51′N, 112°20′E	5.6	2.4

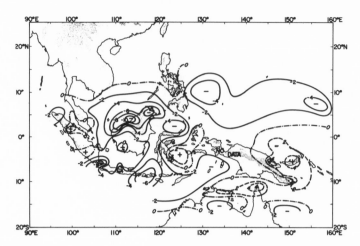

Fig. 7.3. Monthly rainfall. The change from January 1963 to January 1964 is in decimeters.

However, it is the 200-mb temperatures and pressure (Fig. 7.4) that suggest important details of interaction between tropics and higher latitudes. Height gradients were larger in 1963 than in 1964, conforming to the stronger circulation.

North of 20S, 200-mb temperatures were higher and heights were greater in 1964. In the tropics the largest positive differences, occurring over the maritime continent, coincide with the largest negative differences in latent heat release. This apparent paradox demands explanation.

In *January 1963,* some " cause," probably of middle-latitude origin, linked heat source over the maritime continent and heat sink in the region of the subtropical jet stream so effectively that despite the heavy rains no heat accumulated over the maritime continent. Consequently, lapse rates continued to favor deep clouds and rain.

Presumably the vigorous circulation embodied relatively strong surface northerlies, which in turn evaporated relatively large amounts of water from the ocean to maintain heavy rains over the maritime continent. Energy deriving from the rains then acted to keep the circulation vigorous. The heat source and sink were relatively close, connected by a vertical circulation predominantly in the *meridional* plane, upward over the maritime continent, downward south of the jet.

The jet stream, displaced southward and exceptionally strong was hydrodynamically unstable on its southern side. As a possible consequence, the intense blocking ridge, extending from Alaska southward, persisted throughout January 1963. Conceivably, then, energy exported from the maritime continent significantly contributed to maintaining the block, which first

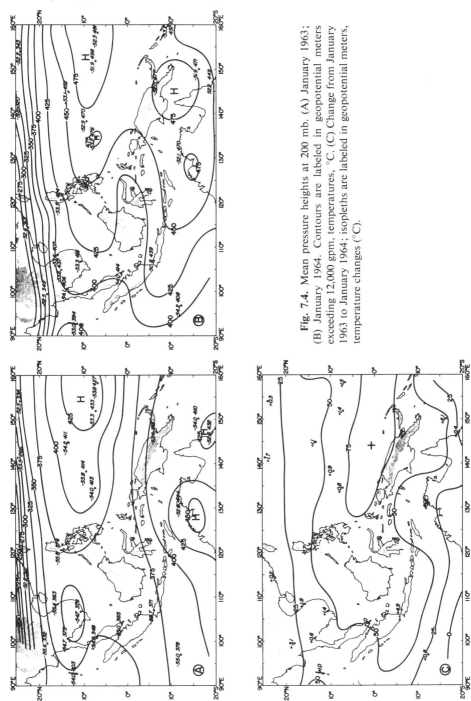

Fig. 7.4. Mean pressure heights at 200 mb. (A) January 1963; (B) January 1964. Contours are labeled in geopotential meters exceeding 12,000 gpm, temperatures, °C. (C) Change from January 1963 to January 1964; isopleths are labeled in geopotential meters, temperature changes (°C).

developed early in the winter above an anomalously warm sea surface (Namias, 1963).

In *January 1964*, the movement of heat from source to extratropical sink was severely restricted. Thus, despite the drought over the maritime continent, released latent heat accumulated in the high troposphere. Lapse rates were less than in January 1963, and therefore not so favorable to convection. Nevertheless, they still exceeded the saturation adiabatic lapse rate. The feeble circulation embodied relatively weak surface northerlies, which by evaporating relatively small amounts of water from the ocean could have further inhibited rains over the maritime continent.

The vertical circulation probably included a significant component in the zonal plane, upward over New Guinea and downward over central Indonesia. This component would be ineffective in dispersing surplus heat or in reinvigorating the horizontal circulation.

Over the maritime continent, 200-mb temperatures were from 1 to 2C higher in January 1964 than in January 1963 (Fig. 7.4c). Let us allow the unlikely explanation that the difference could be wholly accounted for by much greater radiational cooling in January 1963. Even then, an assumed average precipitation excess of 200 mm in that month (Fig. 7.3) implies that at least 400-langleys day^{-1} heat excess over January 1964 was available to fuel the circulation. That this excess was supplied from the ocean is shown by the fact that surface temperatures over the China Seas averaged 1.5C lower in January 1963 than in January 1964.

The difference between January 1963 and January 1964 extended far beyond the monsoon area, where perhaps Grosswetterlagen are triggered by events in middle latitudes. Intensity and persistence of a new regime might be influenced by the response from low latitudes. I am now less puzzled by the considerable weather fluctuations over the maritime continent between one winter and another or within a single winter, despite the fact that tropical storms or other moving disturbances are extremely rare (Section 5.5.3).

Lack of adequate data forced me to ignore the Southern Hemisphere. Although the omission is likely to be unimportant because mean charts show the meridional circulation to be largely confined to the Northern Hemisphere, in some years the Southern Hemisphere could be significantly involved.

7.2.2 FLUCTUATIONS OF THE EAST ASIAN HADLEY CELL IN OTHER WINTERS

The surface-pressure gradient nearly parallels the line between Hong Kong and Djakarta. Thus pressure difference between these two points could provide a rough long-term measure of Hadley cell intensity. Six

Table 7.2

Correlations, based on 50 yr of monthly data, between Hong Kong–Djakarta mean sea-level pressure difference and the average of rainfall at six Indonesian stations[a]

	Nov.	Dec.	Jan.	Feb.	Mar.
Correlation coefficient	0.43	0.50	0.52	0.70	0.65
Significance level (%)		5	5	1	1

[a] Stations: Manado (1°32′N, 124°55′E); Kuching (1°29′N, 110°20′E); Pontianak (0°00′N, 109°20′E); Djakarta (6°11′S, 106°50′E); Pasuruan (7°38′S, 112°55′E); Kupang (10°10′S, 123°40′E).

stations in Indonesia have correspondingly long-term rainfall records (Table 7.2). Since pressure difference and rainfall are positively correlated, the implications of the January 1963 and January 1964 comparison would appear to be worth pursuing. Although the Hong Kong–Djakarta pressure difference roughly measures strength of the northerlies, orographic effects cannot account for the variation in rainfall over Indonesia, for only one of the six stations, Kuching, is exposed to the surface monsoon.

Depperman's (1941) theme recurs—rainfall depends on the magnitude and sign of general vertical motion; orography is a modulator (Sections 5.5.2, 5.8.2.1, 5.8.7.5, and 5.9.2).

7.2.3   FEBRUARY 1952 AND FEBRUARY 1953 OVER NORTHERN AUSTRALIA

February 1952, in the middle of the worst northern Australian drought in 60 years, differed sharply from February 1953, when rains were copious. Bond (1960) analyzed Grosswetterlagen during the two summers and concluded that monsoon *rains* do not set in until the surface pressure difference between Singapore and Darwin exceeds 6 mb, while a break in the *rains* is heralded by the pressure difference decreasing to less than 6 mb.

In February 1952 no tropical cyclones developed off northern Australia and highs migrated eastward across the continent 5° closer to the equator than normal. In February 1953 most of the rain came from two tropical cyclones, while in this month the track of the migrating highs lay up to 10° poleward of normal.

Were the Singapore–Darwin pressure difference determined solely by the strength of the Southeast Asian Hadley cell, then February 1952 should also have been dry over Indonesia. However, all six stations listed in Table 7.2 recorded more rain in February 1952 than in February 1953, while the Hong Kong–Djakarta surface-pressure difference was greater than normal in both years.

When the Northern Hemisphere Hadley cell is strong, pressures are rela-tively low over Indonesia. Flow across the equator into Australia is apparently less than normal, tending to reduce the rainfall over northern Australia as in January 1963 (Fig. 7.3). This could be partially compensated by an increased upward motion associated with a Southern Hemisphere Hadley cell, for the existence of which Berson (1961) found some evidence. In 1951 to 1952, however, high pressure persisting far to the north over Australia kept the heat trough north of the continent, which then experienced the subsidence more typical of winter.

Bond concluded that *coincidence* (fortunately rare) of Northern and Southern Hemisphere rain inhibitors was responsible for this extreme drought. An intriguing question remains—why did the drought last six months?

7.3   Variability of July Rain in Central China

In a normal summer, the Mei-Yü of the Yangtze Valley ceases by mid-July as the polar front weakens and shifts north (Section 5.8.6.3). In some years, however, the shift occurs by the end of June and the valley suffers a drought; in other years when the shift is delayed until the end of July, the month is noteworthy for disastrous floods (Lee and Wan, 1955). In all of China, July rainfall in the Yangtze Valley is the most precarious (Chu, 1962), with variability exceeding 60%. At Hankow (30°35'N, 114°17'E), slight or moderate droughts occurred during 28% of Julys and slight or moderate floods during 14%.

The flood of July 1931 stemmed from a prolonged severe Siberian winter (Lee, 1932). This delayed the ice melt over the Bering and Okhotsk seas, thus maintaining the Okhotsk high for longer than usual. Over southern China warm moist air swept in from the Pacific high to meet the cold polar air along the active polar front. Lows developed, stagnated, and precipitated. The 1954 floods arose from similar causes (Chen, 1957).

In drought years the Sea of Okhotsk warms rapidly and the Okhotsk high becomes weak by the end of June. In flood years, ice melt is delayed over the Bering and Okhotsk seas and the Okhotsk high persists for most of July.

Air–sea interaction as a factor in Grosswetterlagen was first suggested by Helland-Hansen and Nansen (1920) and later exemplified in many case studies by Namias (1963, 1969) and others. The postulated sequence (Section 5.8.4.1) starts with a persistent meteorological regime (in this case, the Siberian anticyclone) which modifies the sea-surface temperature (extent and depth of ice on the Sea of Okhotsk). The enormous heat capacity locked in the sea-surface layers ensures that a temperature anomaly, once produced, tends to persist for weeks or months after the initiating meteorological circulation has disappeared. Thus the Okhotsk high develops in early summer as the land is being rapidly warmed and persists in proportion to the amount of cold originally stored in the winter ice.

Is it possible that the blocking action of the Okhotsk high also affects the character of the South Asian monsoon (Asakura, 1968) (Section 5.8.6.2)?

7.4 Long-Range Forecasting

Long-range forecasts made in the monsoon area generally apply to periods of from a month to a season and are designed to anticipate extreme conditions, particularly floods and droughts.

In analogy to short-period forecasting (Section 6.2), two methods of long-range forecasting appear feasible—statistical or dynamic. I shall first discuss the former, the only method to be extensively used in the monsoon area, and then comment on prospects for the latter.

7.4.1 LONG-RANGE FORECASTING IN INDIA

7.4.1.1 *Early Work.* H. F. Blanford, Meteorological Reporter to the Government of India, made the first unofficial forecast of summer monsoon rainfall in 1881. He based it and subsequent forecasts on the hypothesis that "varying extent and thickness of the Himalayan snows exercise a great and prolonged influence on the climatic conditions and weather of the plains of northwest India" (Blanford, 1884). In 1885 Blanford correctly warned of deficient summer monsoon rain over western India on the basis of late and excessive spring snowfall on the western Himalayas.

The forecasts were then made official. Eliot, Blanford's successor, added other predictors, applying them by means of analogs and parallel curves to the making of increasingly detailed forecasts. However, accuracy did not increase. The disastrously wrong forecasts preceding and during the terrible western Indian drought of 1899 threw the methods into disrepute.

7.4.1.2 *Multiple-Predictor Statistics.* The concept of correlation to test whether preceding events almost anywhere in the world might have a significant relationship with subsequent seasonal rainfall over various Indian climatological subdivisions was introduced by Walker and made the basis of a new method of seasonal forecasting (Walker, 1923). From thousands of computed predictor–predictand correlation pairs, those having the highest correlation coefficients were combined in linear regression equations designed to forecast subsequent seasonal (3-month) rainfall over subdivisions of India.

At the beginning, Walker (1910–1916) warned that the probable error in the highest of a group of correlation coefficients is significantly greater than the probable error of a coefficient selected at random from the group. For example, in a set of 50 correlation pairs ($n = 50$), the respective probable errors are 0.35 and 0.10. In the long and vigorous search for better predictors, Walker's disciples and perhaps even Walker himself appeared at times to ignore this important limitation of the method.

The linear regression equation used by the India Meteorological Department in 1960 (Jagannathan and Khandekar, 1962) to forecast the summer rainfall of peninsular India is typical:

monsoon rain peninsula
$$
\begin{aligned}
= &-0.067 \text{ (Southern Rhodesian rain October to April)} \\
&+1.825 \text{ (South American pressure } \tfrac{1}{2}\text{ (April + May))} \\
&-0.0183 \text{ (Java rain October to February)} \\
&+0.912 \text{ (Bangalore (12°57'N, 77°38'E) northerly winds April)} \\
&-0.559 \text{ (Calcutta (22°39'N, 88°27'E) easterly winds May)} \\
&-4.307
\end{aligned}
$$

Of the predictors used by Walker (1924) in his regression equation only three remained and the correlations of these varied widely through the years (Table 7.3).

Table 7.3

Decade correlation coefficients of the three most consistent predictors of summer monsoon rain over peninsular India[a]

Period	Southern Rhodesian rain	South American pressure	Java rain
1881 to 1890		0.48	−0.11
1891 to 1900		0.78	−0.68
1901 to 1910		0.42	−0.48
1911 to 1920	−0.72	0.27	−0.19
1921 to 1930	−0.24	0.02	−0.24
1931 to 1940	−0.19	0.50	−0.07
1941 to 1950	0.31	0.58	0.04
1951 to 1960		−0.73	−0.11
Entire period	−0.31	0.34	−0.21

[a] From Rao (1965).

Normand (1953) summarized the present position:

"Indian monsoon rainfall has its connections with later rather than with earlier events. The Indian monsoon therefore stands out as an active, not a passive, feature in world weather, more efficient as a broadcasting tool than as an event to be forecast . . . On the whole Walker's worldwide survey ended by offering more promise for the prediction of events in other regions than in India . . . The most important need in monsoon forecasting is to pick out with a reasonable degree of success the years of low rainfall, the possible famine years . . . To ask a regression formula to select the extreme bad years is a very stringent test, which, it must be admitted, the actual forecasts have failed to pass in these twenty years [1933–1952]."

In recent years, Indian meteorologists have been examining radiosonde data for possible inclusion in regression equations on which the seasonal forecasts might be based (Jagannathan and Khandekar, 1962). Significantly, they are considering only data from Indian stations. Walker's attempts to relate events across the world with subsequent rainfall over India were based on tenuous physical justification, and met with limited success.

The immense effects of direct solar heating of the deserts and of latent heat release from the great premonsoon thunderstorms of the Indian subcontinent on the atmosphere over India lead one to think that future monsoon rains may well be more closely related to previous *local* rather than far-off events. Certainly this new approach is worth a fair trial although, of course, correlation methods demand long periods of record for development of regression equations and their subsequent evaluation. As yet, extensive homogeneous upper-air data are available for only the past 20 years, a relatively short period in statistics.

7.4.2 LONG-RANGE FORECASTING IN OTHER MONSOON REGIONS

The apparent success of Walker's early forecasts encouraged emulation (Weightman, 1941). Regression equations were developed by Walker's co-workers for predicting the Nile flood (Bliss, 1926) and Thailand rainfall (Iyer, 1931). Information on performance is lacking.

Under Walker's tutelage Tu (1934) devised similar regression equations for Chinese summer rainfall. Later, however (1937), he concluded that further work had not been worthwhile and presumably made no attempt to apply his results.

Euwe (1949) and de Boer and Euwe (1949) developed regression equations for forecasting seasonal rains over central Indonesia. However, discouraging results of tests on independent data caused them to abandon the method.

The shortcomings of short-period statistical forecast methods revealed in Section 6.2 become accentuated in long-range forecasting. Correlations can seldom be given unequivocal physical interpretations; the all-important extremes are never predicted. I admire the optimists who are sure that the perfect correlations merely await discovery—they may be right. Sea temperature as a predictor has seldom been tested by Walker's followers. Do long records exist of ice extent and thickness during spring in the Sea of Okhotsk (Section 7.3)?

7.4.3   Long-Range Forecasting by Numerical Methods

This has not been attempted specifically for the monsoon area, but since success must embrace the whole atmosphere, the experiments now being conducted or planned incorporate monsoon area forecasting.

The most ambitious model (nine-level, hemispheric) developed by Smagorinsky and his co-workers (Miyakoda *et al.*, 1969) was used to make two two-week wintertime predictions. The effects of orography, radiative transfer, the turbulent exchanges of heat and moisture with the lower boundary, and the difference in thermal properties between the land–ice and sea–ice surfaces were all included; condensation was specified to occur when the relative humidity rose to 80%. The authors found the two-week forecasts "promising" and postulated that more data and a reduced data mesh size would lead to improvements. However, tropical rainfall, particularly over Africa and the central Pacific, was seriously overestimated.

Practitioners of extended-range numerical weather forecasting think that, "Predictions of statistical quantities, such as departures of monthly mean temperatures from normal values, may conceivably be made for longer periods [than two weeks]" (U.S. Committee for GARP, 1969).* Others (Robinson, 1967; Namias, 1968) doubt whether the turbulent atmosphere is deterministically predictable beyond a week or that departures of means from normal values can be accurately estimated beyond two weeks.

Only a detailed global experiment is likely to settle this argument and thereby to decide whether useful long-range numerical forecasts can be achieved for the monsoon area.

---

* The Committee had the middle latitudes in mind; its optimism did not include the tropics.

Better is the end of a thing than the beginning thereof.

Ecclesiastes, VII; 8.

8. Concluding Remarks

In preceding chapters I have tried to demonstrate that monsoon-area climate and weather comprise a peculiar regional subdivision of the atmospheric sciences. The area is subjected to an annual cycle in ocean–continent temperature gradient. During summer and winter, although surface disturbances are rare, middle- or upper-tropospheric disturbances are not. In the transition seasons differences between the monsoon area and surrounding areas are least. Thus latitude for latitude, the range and complexity of annual variations are greater in the monsoon area than beyond.

The general character of the monsoons and their interregional variations reflect the juxtaposition of continents and oceans and the presence or absence of upwelling. However, without the great mechanical and thermal distortions produced by the Himalayas and the Tibetan Plateau, the vast Northern Hemisphere deserts would be less desertlike, central China would be much drier and no colder in winter than India, while even over the Coral Sea winter cloud and rain would be uncommon.

Within the monsoon area annual variations are seldom spatially or temporally in phase. Even if these variations were understood and their phases successfully forecast, accurate day-to-day weather prediction would not necessarily be achieved, for the climatological cycles merely determine *necessary* conditions for certain weather regimes; synoptic changes then control where and when the rain will fall and how heavily, and whether winds will be destructive.

Although not new, perhaps the most important concept to be developed in this book is that of wide-ranging, nearly simultaneous accelerations or decelerations within a major vertical circulation. Causes are elusive, although the changes generally appear to be triggered by prior changes in the heat-sink regions of the vertical circulation. This is a field of truly enormous potential for numerical modeling, on a time scale between synoptic and seasonal, in which fluctuations in radiation and in air–surface energy exchange might produce profound effects.

The concept both explains previous difficulty in maintaining continuity of synoptic analysis, and demands that notions of day-to-day weather changes be examined and probably modified. Even during winter, fronts seldom remain material boundaries for long and air-mass analysis confuses more often than not.

That synoptic-scale disturbances often appear to develop and to weaken in response to changes in the major vertical circulations might explain why many of the disturbances are quasi stationary. In turn, synoptic-scale vertical motion determines the character of convection and the efficiency with which energy is transported upward from the heat source.

Synoptic-scale lifting, by spreading moisture deeply through the troposphere, reduces the lapse rate and increases the heat content in mid-troposphere. Thus, though it diminishes the intensity of small-scale convection and the frequency of thunderstorms, it increases rainfall and upward heat transport. Conversely, synoptic-scale *sinking*, by drying the mid-troposphere, creates a heat minimum there, hinders upward transport of heat and diminishes rainfall. However, the increased lapse rate favors scattered, intense, small-scale convection and thunderstorms.

In the monsoon area the character of the weather, on the scale of individual clouds, seems to be determined by changes occurring successively on the macro- and synoptic scales. *Rains* set in—not when cumulonimbus gradually merge—but when a synoptic disturbance develops, perhaps in response to change in a major vertical circulation. *Showers* too are part of the synoptic cycle. Individually intense, but collectively less wet, they succeed or precede *rains* as general upward motion diminishes.

When synoptic-scale lifting is combined with very efficient upper-tropospheric heat disposal, the lapse rate may be steep enough to support intense convection. Then a vast continuous thunderstorm gives prolonged torrential rain. Many times this takes place within the common upward branch of two major vertical circulations.

I have omitted topics which might have been included: micrometeorology and cloud physics because peculiarly monsoon characteristics have not yet been identified; and monsoons in the stratosphere because they are of global extent and are caused by annual photochemical as well as radiational variations. Because the results resist rigorous physical interpretation I have reluctantly desisted from trying to describe and evaluate the statistical contributions of eddies to the total energy of monsoon circulations (c.f. Riehl, 1954; Flohn, 1964a).

Many tropical meteorologists, enslaved by meteorological fashion, have striven to make their work appear as quantitative and objective as possible.

This commendable aim has led to important *climatological* insights. However, in synoptic studies their quantitative results have usually been belied by Nature's quantities. A numerical model which determines that air is massively rising over the deserts of Arabia has limited validity, no matter how quantitative and objective it might be. Energy-budget computations in which precipitation and evaporation are residuals, or must be estimated, have also had their day.

I therefore make no apologies for my largely qualitative approach. Without adequate descriptions in their diets, computers suffer malnutrition and spawn monsters. Aviators, not theoreticians, discovered the jet stream.

Forecasting and research should be inseparable. The very few monsoon-area weather services which enable their forecast meteorologists to spend at least one-third of their time on research have thereby greatly enhanced staff morale and their scientific reputations, to say nothing of improved forecast accuracy.

Combined forecast research programs could well be successfully directed to solving some of the problems discussed in preceding chapters and to increasing the number of recognizable models of synoptic circulations.

The area covered by synoptic analyses should be sufficiently broad for the major vertical circulations to be monitored. Then interaction with synoptic disturbances and consequent effects on rainfall could be detected and possibly anticipated.

Over the Sahara, information from good APT reception and an adequate surface observing network might be applied to the fascinating problems of the role of dust in air-mass discontinuities and of the origin and movement of haboobs.

Mesoscale gradients within synoptic systems and their diurnal variations might better be understood were studies to combine information from weather radars and weather satellites. Ceraunograms (Southern *et al.*, 1970) could help bridge the gap between synoptic and mesoscales. Aerial probing of continuous thunderstorms would likely illuminate the shadowy picture we now have of energy transformations.

We should view the future of monsoon meteorology with optimistic discontent. Regional progress in understanding and forecasting weather has been disappointingly slow. But I hope I have demonstrated that attacks are being vigorously pressed on problems of concern to the entire monsoon area. I urge enthusiastic *intraarea* exchange of people and ideas. Conceivably, Nairobi and Poona meteorologists exchanging visits makes better sense than their spending time at great computing centers where they may learn to perform minor cosmetic surgery on the latest nine-level primitive-equation general circulation model.

Plate 1. Mean conditions during January.

PLATE I   MEAN CONDITIONS DURING JANUARY   247

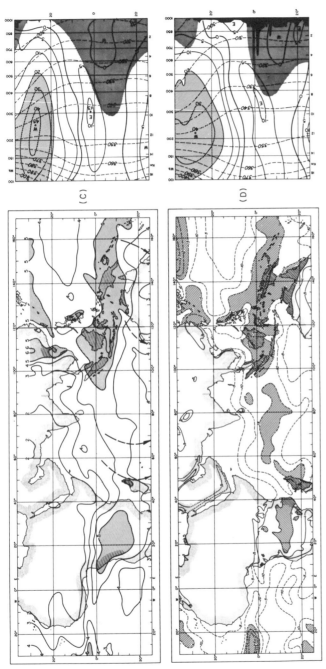

Left-hand diagrams: (A) Resultant 200-mb flow (isotachs in meters per second) (from Ramage and Raman, 1970; Frost, 1969; Wiederanders, 1961). (B) Resultant surface flow (isotachs in Beaufort force, shown only over the oceans) (from van Duijnen Montijn, 1952; Meteorological Office, 1947, 1949; Tao, 1949; Thompson, 1965; Taylor, 1932). (C) Cloudiness in oktas based on 3 yr of weather satellite pictures. Heavy dashed arrows show predominant tropical cyclone tracks. Areas with a high frequency of tropical cyclones are stippled. (D) Rainfall over land in decimeters (full lines) (from Braak, 1921–1929; Taylor, 1932; Chu, 1936; Biel, 1945; Ramanathan and Venkiteshwaran, 1948; Thompson, 1965); percentage frequency of ship observations reporting precipitation (dashed lines) (from U.S. Weather Bureau, 1955–1959).

Right-hand diagrams: Meridional profiles of (1) zonal wind components in meters per second (full lines), (2) potential temperature in degrees Kelvin (dashed lines), and (3) layers with relative humidities greater than 50% (stippled) (from Ramage and Raman, 1970). (A) 140E; (B) 100E; (C) 73E; (D) 30E.

PLATE II   MEAN CONDITIONS DURING APRIL   249

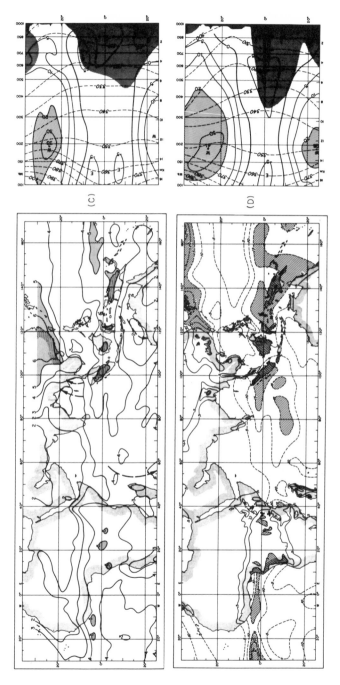

Plate II. Mean conditions during April (data measurements taken same as those in Plate I).

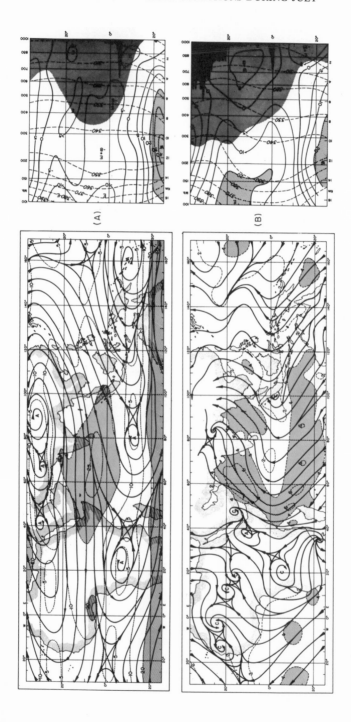

PLATE III   MEAN CONDITIONS DURING JULY   251

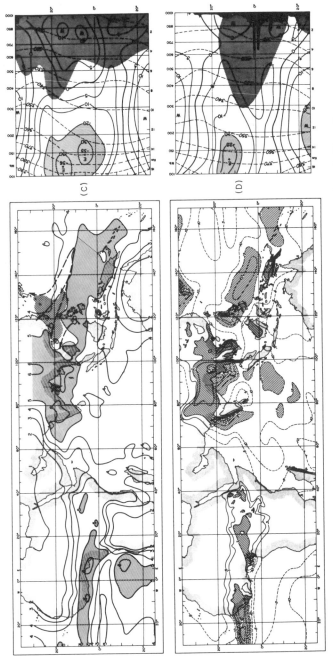

Plate III. Mean conditions during July (data measurements taken same as those in Plate I).

PLATE IV   MEAN CONDITIONS DURING OCTOBER   253

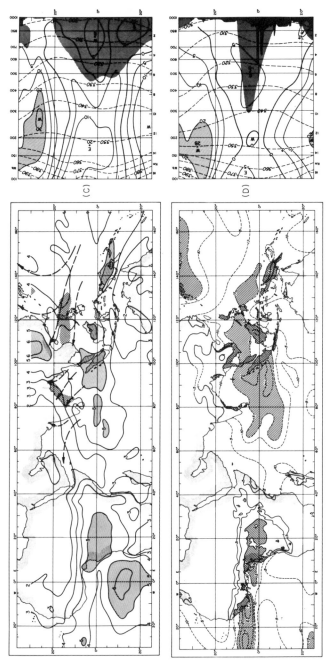

Plate IV. Mean conditions during October (data measurements taken same as those in Plate I).

List of Symbols

Dimension symbols follow identification: L = length; T = time; M = mass; Θ = temperature. Cross section references are found in the right-hand column.

a	radius of earth $= 6.37 \times 10^6$ m (mean)	L	3.10.1
A	area	L^2	3.10.1.2
A_ζ	vorticity advection $= -\mathbf{V} \cdot \nabla \zeta$, at level of nondivergence	T^{-2}	3.1; 3.2
A_T	thickness advection between 1000 mb (Z_{1000}) and level of nondivergence (Z); $= -\mathbf{V} \cdot \nabla(Z - Z_{1000})$	LT^{-1}	3.1; 3.2
C	in-cloud horizontal vector wind	LT^{-1}	3.11
C_D	drag coefficient $= \tau_0/(\rho \overline{V}^2)$		2.2.2; 3.8.2
c_p	specific heat of dry air at constant pressure $= 7R/2 = 0.24$ cal gm^{-1} K^{-1}	$L^2T^{-2}\Theta^{-1}$	3.1; 3.2; 3.5.2
d	distance	L	6.1.4
D	diameter	L	3.11; 6.1.4
d/dt	differentiation following the motion	T^{-1}	
e_a	water vapor pressure of air at 10 m above the surface	$L^{-1}MT^{-2}$	2.4.1
e_s	saturation water vapor pressure at the sea-surface temperature	$L^{-1}MT^{-2}$	2.4.1
f	Coriolis parameter $= 2\Omega \sin \phi$	T^{-1}	
F	coefficient of surface resistance	T^{-1}	3.10.1.2
g	acceleration of gravity ≈ 980 cm sec^{-2}	LT^{-2}	
h	pressure height	L^2T^{-2}	6.2.1.1
$\mathbf{i}$	unit vector in X direction		
H	heat per unit mass	L^2T^{-2}	3.1; 3.2; 3.8.1
$\mathbf{j}$	unit vector in Y direction		
J	Jacobian determinant		6.2.1.1
k	correction factor for cloud cover from Roden (1959)		2.4.1
k'	empirical factor from Roden (1959) based on average cloud cover and sun's altitude, for selected latitudes		2.4.1
k''	empirical factor from Roden (1959) which is a function of latitude and the climatological vertical extent of cloud cover		2.4.1
$\mathbf{K}$	vector wind correction	LT^{-1}	6.1.4

L	latent heat of vaporization of water ≈ 600 cal gm^{-1}	L^2T^{-2}	3.5.2
n	number of pairs used in determining the correlation between two series of quantities		
N	cloudiness in percent		5.5
$\bar{N}$	average total cloud cover in tenths		2.4.1
p	atmospheric pressure	$L^{-1}MT^{-2}$	
q	specific humidity		3.5.2
Q_0	rate of heating at the surface from incoming short-wave radiation under clear skies	MT^{-3}	2.4.1
Q_b	rate of emission of back radiation from the surface	MT^{-3}	2.4.1
Q_e	rate of cooling at the surface from evaporation	MT^{-3}	2.4.1
Q_h	rate of sensible heat loss from the surface to the atmosphere	MT^{-3}	2.4.1
Q_r	rate at which radiation is reflected at the surface	MT^{-3}	2.4.1
Q_s	rate of heating at the surface from incoming short-wave radiation, modified by the total average cloud cover (insolation)	MT^{-3}	2.4.1
Q_{ST}	net heat balance at the surface	MT^{-3}	2.4.1; 5.5
r	correlation coefficient between two series of quantities		
r_t	correlation coefficient between two series of quantities separated by a time interval t		6.2.2.2
R	gas constant for dry air $= 6.8557 \times 10^{-2}$ cal gm^{-1} K^{-1}	$L^2T^{-2}\Theta^{-1}$	
s	surface wind speed	LT^{-1}	2.2.2
S	emissivity of the sea ≈ 0.98		2.4.1
t	time	T	
T	temperature	Θ	
T_a	temperature of the air at 10 m above surface	Θ	2.4.1
T_e	temperature of environment	Θ	3.11.1
T_s	temperature of sea surface	Θ	2.4.1; 5.5
T_v	virtual temperature	Θ	4.1.1; 5.7
u	wind velocity component in X direction or in the zonal direction (W $\to$ E)	LT^{-1}	
u_g	geostrophic wind velocity component in X direction or in the zonal direction	LT^{-1}	6.1.1.2
v	wind velocity component in Y direction or in the meridional direction (S $\to$ N)	LT^{-1}	

V	horizontal wind	LT^{-1}	
$\mathbf{V}$	horizontal vector wind	LT^{-1}	
$\mathbf{V}_e$	environmental horizontal vector wind	LT^{-1}	3.11.2
V_g	geostrophic wind	LT^{-1}	6.2.1.1
V_{g_0}	surface geostrophic wind	LT^{-1}	3.8.2
$\mathbf{V}_r$	relative horizontal vector wind	LT^{-1}	3.11
V_ψ	stream function wind	LT^{-1}	6.2.1.1
w	wind velocity component in Z direction or vertically	LT^{-1}	
∇	two-dimensional Hamilton operator $=\mathbf{i}\,\partial/\partial x +\mathbf{j}\,\partial/\partial y$	L^{-1}	
$\nabla\cdot\mathbf{V}$	divergence of horizontal wind $= \partial u/\partial x + \partial v/\partial y$	T^{-1}	
∇^2	two-dimensional Laplacian operator $= \partial^2/\partial x^2 + \partial^2/\partial y^2$	L^{-2}	
β	Rossby parameter $= 2\Omega \cos \phi/a$	$L^{-1}T^{-1}$	3.1
γ	lapse rate $= -dT/dz$	$L^{-1}\Theta$	3.8.2
γ_a	dry adiabatic lapse rate $= g/c_p$	$L^{-1}\Theta$	
γ_s	saturated adiabatic lapse rate $= g[2.39 \times 10^{-8} + 0.621(eLpRT)]/[c_p + 0.621(L\,de/p\,dT)]$	$L^{-1}\Theta$	3.8.2
Γ	lapse rate in terms of pressure $= \gamma/\rho g$	$LM^{-1}T^2\Theta$	3.1; 3.2; 3.8.1
Γ_a	dry adiabatic lapse rate in terms of pressure $= \gamma_a/\rho g$	$LM^{-1}T^2\Theta$	3.1; 3.2; 3.8.1
ζ	component of relative vorticity normal to a specified surface. ζ normal to an x–y surface $= \partial v/\partial x - \partial u/\partial y$	T^{-1}	6.2.1.1
ζ_0	ζ at earth's surface	T^{-1}	3.1; 3.2
ζ_a	component of absolute vorticity normal to a specified surface $= f + \zeta$	T^{-1}	3.5.4.1; 6.2.1.1
ζ_g	geostrophic vorticity	T^{-1}	6.2.1.1
ζ_F	vorticity of the frictional force	T^{-2}	3.5.4.1
θ	potential temperature	Θ	
θ_e	equivalent potential temperature	Θ	3.5.2
θ_{ve}	virtual equivalent potential temperature	Θ	3.5.2; 4.1.1; 5.8.7.5
λ	angle of longitude		
μ	relative humidity ($1 = 100\%$)		3.8.2
ρ	air density	$L^{-3}M$	
σ	Stefan–Boltzman constant $= 1.355 \times 10^{-12}$ cal cm^{-2} K^{-4} sec^{-1}	$MT^{-3}\Theta^{-4}$	2.4.1
τ	turbulent shear stress	$L^{-1}MT^{-2}$	2.2.1
τ_0	turbulent shear stress at the surface	$L^{-1}MT^{-2}$	2.2.1
ϕ	angle of latitude		
ψ	stream function $u = -\partial\psi/\partial y$; $v = \partial\psi/\partial x$ and $\partial u/\partial x + \partial v/\partial y = 0$	L^2T^{-1}	6.2.1.1
ω	vertical velocity with pressure as vertical coordinate $= dp/dt$	$L^{-1}MT^{-3}$	3.1; 3.2; 6.1.5
$\tilde{\omega}$	nonhydrostatic pressure	$L^{-1}MT^{-2}$	3.11
Ω	angular velocity of earth's rotation	T^{-1}	

Bibliography

Readers will find the following general reference useful:

NG, N. H. (1967). Acta Meteorologica Sinica, translated titles and abstracts from Vol. 27 (1956) through Vol. 35 (1965), Transl. Emm–67–167. Oriental Science Library, Emmanuel College, Boston, Massachusetts.

References

ABLES, C. D. (1969). The Structure of a Heat Low. M.S. Thesis, Dept. of Geosciences, Univ. of Hawaii, Honolulu.

AGARWAL, O. P., AND KRISHNAMURTHY, G. (1969). A radar study of line-type thunderstorms over Bombay airport and surroundings. *Indian J. Meteorol. Geophys.* **20**, 36–40.

ALISOV, B. P. (1954). "Die Klimat der Erde." Deutscher Verlag, Berlin.

ANANTHAKRISHNAN, R. (1964). Tracks of Storms and Depressions in the Bay of Bengal and the Arabian Sea 1877–1960. India Meteorol. Dept., New Delhi.

ANANTHAKRISHNAN, R., (1966). Monsoon Rainfall Summary, June to September, 1966. India Meteorol. Dept., Poona.

ANANTHAKRISHNAN, R., AND BHATIA, K. L. (1960). Tracks of monsoon depressions and their recurvature towards Kashmir. *In* "Monsoons of the World" (S. Basu *et al.*, eds.), pp. 157–172. Hind Union Press, New Delhi.

ANANTHAKRISHNAN, R., and KESHAVAMURTHY, R. N. (1970). On some aspects of the fluctuations in the pressure and wind fields over India during the summer and winter monsoon seasons. *In Proc. Symp. Tropical Meteorol. Honolulu*, pp. L III–1–6. Amer. Meteorol. Soc., Boston, Massachusetts.

ANANTHAKRISHNAN, R., and PATHAN, J. M. (1970). North-south oscillations of the equatorial trough and seasonal variations of rainfall in the tropics. *In Proc. Symp. Tropical Meteorol. Honolulu*, pp. F V–1–5. Amer. Meteorol. Soc., Boston, Massachusetts.

ANANTHAKRISHNAN, R., AND RAO, K. V. (1964). Diurnal variation of low level circulation over India. *In Proc. Symp. Tropical Meteorol.* (J. W. Hutchings, ed.), pp. 89–95. New Zealand Meteorol. Serv., Wellington.

ANANTHAKRISHNAN, R., AND THIRUVENGADATHAN, A. (1967). Comparison between the observed and geostrophic winds over the Indian tropics during the summer and winter monsoons. *In Proc. Symp. Meteorol. Results Int. Indian Ocean Expedition* (P. R. Pisharoty, ed.), pp. 384–397. India Meteorol. Dept., New Delhi.

ANDERSON, R. K., FERGUSON, E. W., AND OLIVER, V. J. (1966). The use of satellite pictures in weather analysis and forecasting. *W.M.O. Tech. Note No.* 75.

ANDREWS, J. F. (1964). The weather and circulation of January 1964. *Mon. Weather Rev.* **92**, 189–194.

ANJANEYULU, T. S. S. (1969). On the estimates of heat and moisture over the Indian monsoon trough zone. *Tellus* **21**, 64–75.

ANTHES, R. A., AND JOHNSON, D. R. (1968). Generation of available potential energy in Hurricane Hilda (1964). *Mon. Weather Rev.* **96**, 291–302.

ARAKAWA, H. (1963). Studies on Statistical Prediction of Typhoons. Nat. Hurricane Res. Project, Rep. No. 61.

ASAKURA, T. (1968). Dynamic climatology of atmospheric circulation over East Asia centered in Japan. *Pap. Meteorol. Geophys.* **19**, 1–67.

AUBERT, E. J., LUND, I. A., AND THOMASELL, A. (1959). Some objective six-hour predictions prepared by statistical methods. *J. Meteorol.* **16**, 436–446.

BAUR, F. (1951). Extended range weather forecasting. *Compendium Meteorol.* 814–833.

BEDEKAR, V. C., AND BANERJEE, A. K. (1969). A study of climatological and other rainfall patterns over central India. *Indian J. Meteorol. Geophys.* **20**, 23–30.

BEDI, H. S., AND PARTHASARATHY, B. (1967). Cold waves over Northwest India and neighbourhood. *Indian J. Meteorol. Geophys.* **18**, 371–378.

BEDIENT, H. A., AND VEDERMAN, J. (1964). Computer analysis and forecasting in the tropics. *Mon. Weather Rev.* **92**, 565–577.

BEDIENT, H. A., COLLINS, W. G., AND DENT, G. (1967). An operational tropical analysis system. *Mon. Weather Rev.* **95**, 942–949.

BELL, G. J. (1968). Personal communication.

BELL, G. J. (1969). Comments in: The Diagnosis and Prediction of SE Asia Northeast Monsoon Weather. U.S. Navy Weather Res. Facility Staff, Norfolk, Virginia.

BELL, G. J. (1970). The monsoon trough the near South China coast. *In Proc. Conf. Summer Monsoon Southeast Asia* (C. S. Ramage, ed.), pp. 83–106. Navy Weather Res. Facility, Norfolk, Virginia.

BELL, G. J. AND CHIN, P. C. (1968). The probable maximum rainfall in Hong Kong. *Roy. Observ. Hong Kong Tech. Mem. No.* 10.

BELLAMY, J. C. (1949). Objective calculations of divergence, vertical velocity, and vorticity. *Bull. Amer. Meteorol. Soc.* **30**, 45–49.

BENEST, E. E. (1923). Notes on the "Sumatras" of the Malacca Straits. *Quart. J. Roy. Meteorol. Soc.* **49**, 237–238.

BERGERON, T. (1949). The problem of artificial control of rainfall on the globe. II: The coastal orographical maxima of precipitation in autumn and winter. *Tellus* **1**, 3; 15–32.

BERGERON, T. (1954). The problem of tropical hurricanes. *Quart. J. Roy. Meteorol. Soc.* **80**, 131–164.

BERSON, F. A. (1961). Circulation and energy balance in a tropical monsoon. *Tellus* **13**, 472–485.

BHALOTRA, Y. P. R. (1963). Meteorology of Sudan. *Mem. Sudan Meteorol. Serv.* **6**.

BHULLAR, G. S. (1952). Onset of monsoon over Delhi. *Indian J. Meteorol. Geophys.* **3**, 25–30.

BIEL, E. R. (1929). Die Veränderlicheit der Jahressumme des Niederschläge auf der Erde. *Geogr. Jahresber. Oesterr.* **14/15**, 151–180.

BIEL, E. R. (1945). Weather and Climate of China. Weather Div., Headquarters Army Air Forces, Washington, D.C.

BILLA, H. S., AND RAJ, H. (1966). Determination of the horizontal dimensions of mesoscale precipitating system. *Indian J. Meteorol. Geophys.* **17**, 383–384.

BJERKNES, J. (1951). Extratropical cyclones. *Compendium Meteorol.* 577–598.

BJERKNES, V., HESSELBERG, TH., AND DEVIK, O. (1911). Dynamic Meteorology and Hydrography. Carnegie Inst., Washington, D.C.

BLANC, M. L. (1965). Requirements of agriculture for long-range forecasts. In WMO-IUGG Symposium on research and development aspects of long-range forecasting pp. 11–16. *W.M.O. Tech. Note No.* 66.

BLANFORD, H. F. (1884). On the connexion of the Himalaya snowfall and seasons of drought in India. *Proc. Roy. Soc. London* **37**, 3–22.

BLISS, E. W. (1926). The Nile flood and world weather. *Mem. Roy. Meteorol. Soc.* **1**, 79–85.

BLÜTHGEN, J. (1966). " Allgemeine Klimageographie," 2nd ed. de Gruyter, Berlin.

BÖHNECKE, G. (1936). Temperatur, Salzgehalt und Dichte an der Overflache des Atlantischen Ozeans. Deutsche Atlantische Expedition auf dem Forschungs-und Vermessungsschiff "Meteor," 1925–27, Vol. 5, 74 charts. Berlin and Leipzig, 1936.

BOLIN, B. (1950). On the influence of the earth's orography on the general character of the westerlies. *Tellus* **2**, 184–195.

BOLIN, B. (ed.) (1967). Global Atmospheric Research Programme (GARP), Report of the Study Conference held at Stockholm, 28 June–11 July 1967. ICSU/IUGG Committee on Atmospheric Sciences.

BOND, H. G. (1960). The drought of 1951–52 in northern Australia. *In* " Monsoons of the World " (S. Basu *et al.*, eds.), pp. 215–222. Hind Union Press, New Delhi.

BOOTH, A. L., AND TAYLOR, V. R. (1969). Mesoscale archive and computer products of digitized video data from ESSA satellites. *Bull. Amer. Meteorol. Soc.* **50**, 431–438.

BOUCHER, R. J., AND WEXLER, R. (1961). The motion and predictability of precipitation lines. *J. Meteorol.* **18**, 160–171.

BRAAK, C. (1921–29). Het Climaat van Nederlandsch Indie. *Magnet. Meteorol. Observ. Batavia, Verhand. No.* 8.

BRAAK, C. (1940). Over de oorzaken van te tijdelijke en plaatselijke verschillen in den neerslag. *Kon. Ned. Meteorol. Inst. Mededel., Verhand. No.* 45.

BROOKFIELD, H. C., AND HART, D. (1966). Rainfall in the Tropical Southwest Pacific. Aust. Nat. Univ., Dept. of Geography, Publ. G, 3.

BRUNT, A. T., AND MACKERRAS, D. (1961). A study of thunderstorms in Southeast Queensland. *Aust. Meteorol. Mag.* **34**, 15–44.

BRUZON, E., AND CARTON, P. (1930). " Le Climat de l'Indochine et les Typhons de la Mer de Chine." Imprimerie d'Extrême-Orient, Hanoi.

BRYSON, R. A., AND KUHN, P. M. (1961). Stress-differential induced divergence with application to littoral precipitation. *Erdkunde* **15**, 287–294.

BUAJITTI, K. (1964). The use of vertical vorticity of 24-hour wind changes for weather forecasting. *In Proc. Symp. Tropical Meteorol.* (J. W. Hutchings, ed.), pp. 393–402. New Zealand Meteorol. Serv., Wellington.

BUDYKO, M. L. (1956). The Heat Balance of the Earth's Surface. Gidrometeorologicheskoe Izdatel'stvo, Leningrad. [Engl. Transl. (1958). U.S. Weather Bur., Washington, D.C.]

BUNKER, A. F. (1967). Interaction of the summer monsoon air with the Arabian Sea. *In Proc. Symp. Meteorol. Results Int. Indian Ocean Expedition* (P. R. Pisharoty, ed.), pp. 3–16. India Meteorol. Dept., New Delhi.

BUNKER, A. F., AND CHAFFEE, M. (1970). Tropical Indian Ocean clouds. *Int. Indian Ocean Expedition, Meteorol. Monogr.* **4**.

BURKE, C. J. (1945). Transformation of polar continental air to polar maritime air. *J. Meteorol.* **2**, 94–112.

BYERS, H. R. (1959). " General Meteorology." McGraw-Hill, New York.

BYERS, H. R., AND BRAHAM, R. R. (1949). The Thunderstorm. U.S. Government Printing Office, Washington, D.C.

CARLSON, T. N. (1969a). Synoptic histories of three African disturbances that developed into Atlantic hurricanes. *Mon. Weather Rev.* **97**, 256–276.

CARLSON, T. N. (1969b). Some remarks on African disturbances and their progress over the tropical Atlantic. *Mon. Weather Rev.* **97**, 716–726.

CARLSTEAD, E. M. (1967). Tropical Numerical Weather Predictions in Hawaii—A Status Report. U.S. Weather Bur. Tech. Mem. PR-4.

CARLSTEAD, E. M. (1969). Personal communication.

CHAKRAVORTTY, K. C., AND BASU, S. C. (1957). The influence of western disturbances on the weather over northeast India in monsoon months. *Indian J. Meteorol. Geophys.* **8**, 261–272.

CHAO, S. (1965). Monsoons and rainy season in Kweichow (in Chinese). *Acta Meteorol. Sinica* **35**, 96–106.

CHARNEY, J. G. (1967). Some remaining problems in numerical weather prediction. *In* "Advances in Numerical Weather Prediction," pp. 61–70. Travelers Research Center, Hartford, Connecticut.

CHARNEY, J. G., AND ELIASSEN, A. (1964). On the growth of the hurricane depression. *J. Atmos. Sci.* **21**, 68–75.

CHAUDHURY, A. M. (1950). On the vertical distribution of wind and temperature over Indo-Pakistan along meridian 76°E in winter. *Tellus* **2**, 56–62.

CHAUSSARD, A. (1960). Quelques particularités des cyclones tropicaux dans le sud-ouest de l'Océan Indien. *In* "Tropical Meteorology in Africa" (D. J. Bargman, ed.), pp. 226–231. Munitalp Foundation, Nairobi.

CHEN, H.-Y. (1957). The circulation characteristics during the 1954 flooding periods in the Yangtze and Hwai Ho basins. (in Chinese). *Acta Meteorol. Sinica* **28**, 1–12.

CHEN, T. Y. (1969). The Severe Rainstorms in Hong Kong during June 1966. Roy. Observ. Hong Kong, Suppl. Meteorol. Results 1966.

CHIN, P. C. (1958). Tropical cyclones in the western Pacific and the China Sea area. *Roy. Observ. Hong Kong, Tech. Mem. No. 7.*

CHIN, P. C. (1969). Cold surges over South China. *Roy Observ. Hong Kong, Tech. Note No. 28.*

CHIN, P. C., AND LEONG, H. C. (1964). Errors of upper-wind forecasts. *Roy. Observ. Hong Kong, Tech. Note No. 21.*

CHOW, S.-P. AND KOO, C.-C. (1958). The influence of large-scale topography on the propagation of upper-level planetary waves (in Chinese). *Acta Meteorol. Sinica* **29**, 99–103.

CHU, C.-C. (1936). Chinese rainfall. *Acad. Sinica, Nat. Res. Inst. Meteorol. Mem. No. 7.*

CHU, P-H. (1962). "Climate of China." Science Publ. Peiping. [Engl. Transl. (1967). U.S. Dept. of Commerce, Joint Publ. Res. Service.]

CLARKE, R. H. (1962). Severe local wind storms in Australia. *CSIRO Div. Meteorol. Phys. Tech. Pap.* **13.**

COCHEMÉ, J. (1960). Some streamlines and contours over the equator. *In* "Tropical Meteorology in Africa" (D. J. Bargman, ed.), pp. 181–211. Munitalp Foundation, Nairobi.

COCHEMÉ, J., AND FRANQUIN, P. (1967). An agroclimatology survey of a semiarid area in Africa south of the Sahara. *W.M.O. Tech. Note. No. 86.*

COLON, J. A. (1964). On interactions between the southwest monsoon current and the sea surface over the Arabian Sea. *Indian J. Meteorol. Geophys.* **15**, 183–200.

COLON, J. A., AND NIGHTINGALE, W. R. (1963). Development of tropical cyclones in relation to circulation patterns at the 200 millibar level. *Mon. Weather Rev.* **91**, 329–336.

CONRAD, V. (1936). Die Klimatologischen Elemente und ihre Abhängigkeit von Terrestrischen Einflüssen. *In* "Handbuch der Klimatologie" (W. Köppen and R. Geiger, eds.), Vol. 1, Part B. Borntraeger, Berlin.

CREUTZBERG, N. (1950). Klima, klimatypen und klimakarten. *Petermanns Geogr. Mitt.* **94**, 57–69.

CUMING, M. J. (1968). Cool season tropical disturbances in the South China Sea and their effect on the weather of Hong Kong. *Roy. Observ. Hong Kong Tech. Note No. 27.*

DALLAS, W. L. (1887). Memoir on the Winds and Monsoons of the Arabian Sea and North Indian Ocean. Superintendent of Government Printing, India, Calcutta.

DALLAS, W. L. (1900). A discussion on the failure of the southwest monsoon rains in 1899. *Mem. India Meteorol. Dept.* **12**, 1–30.

DALLAS, W. L. (1902). Meteorological history of the seven monsoon seasons, 1893–1899, in relation to the Indian rainfall. *Mem. India Meteorol. Dept.* **12**, 409–486.

DAMPIER, W. (1697). "A New Voyage Round the World." James Knapton, London.

DANIEL, C. E. J., AND SUBRAMANIAM, A. H. (1966). Exceptionally heavy rain in South Kerala on 17/18 October 1964. *Indian J. Meteorol. Geophys.* **17**, 253–256.

DANIELSEN, E. F., AND HO, F. P. (1969). An Isentropic Trajectory Study of a Strong Northeast Monsoon Surge. Hawaii Inst. Geophys. Rep. No. 69–3.

DAO, S.-Y., ZHAO, Y.-J., AND CHEN, X.-M. (1958). The relationship between the Mei-Yü period in the Far East and the seasonal variation of the upper-air circulation over Asia (in Chinese). *Acta Meteorol. Sinica* **29**, 119–134.

DAO, S.-Y., HSÜ, S.-Y., AND KUO, C.-Y. (1962). The characteristics of the zonal and meridional circulation over tropical and subtropical regions in East Asia in summer (in Chinese). *Acta Meteorol. Sinica* **32**, 91–103.

DAS, P. M., DE, A. C., AND GANGOPADHYAYA, M. (1957). Movements of two Nor'westers of West Bengal: A radar study. *Indian J. Meteorol. Geophys.* **8**, 399–406.

DATTA, R. K., AND GUPTA, M. G. (1967). Synoptic study of the formation and movements of western depressions. *Indian J. Meteorol. Geophys.* **18**, 45–50.

DAVIS, R. A. (1969). A Computer Method to generate and plot Streamlines. U.S. Weather Bur. Tech. Mem. No. PR-5.

DE, A. C. (1963). Movement of pre-monsoon squall lines over gangetic West Bengal as observed by radar at Dum Dum Airport. *Indian J. Meteorol. Geophys.* **14**, 37–45.

DE, A. C., DAS, P. M., AND GANGOPADHYAYA, M. (1957). Regenerative drift of a thunderstorm squall of the southwest monsoon season. *Indian J. Meteorol. Geophys.* **8**, 72–80.

DE BOER, H. J., AND EUWE, W. (1949). Forecasting rainfall for the Period July–August–September for parts of Celebes and South Borneo. *Kon. Magnet. Meteorol. Observ. Batavia, Verhand. No.* 36.

DELORMÉ, G. A. (1963). Repartition et durée des précipitations en Afrique Occidentale. *France, Météorol. Nat. Monogr.* **28**.

DEPPERMAN, C. E. (1941). On the occurrence of dry stable maritime air in equatorial regions. *Bull. Amer. Meteorol. Soc.* **22**, 143–149.

DESAI, B. N. (1950). Mechanism of nor'westers in Bengal. *Indian J. Meteorol. Geophys.* **1**, 74–76.

DESAI, B. N. (1951). On the development and structure of monsoon depressions in India. *Mem. India Meteorol. Dept.* **28**, 217–228.

DESAI, B. N., AND KOTESWARAM, P. (1951). Air masses and fronts in the monsoon depressions in India. *Indian J. Meteorol. Geophys.* **2**, 250–265.

Deutsches Hydrographisches Institut (1960). "Monats-karten für den Indischen Ozean." Hamburg.

DIXIT, C. M., AND JONES, D. R. (1965). A Kinematic and Dynamical Study of Active and Weak Monsoon Conditions over India during June and July, 1964. Int. Meteorol. Center Bombay (prepub.).

DUNCAN, M. H. (1930). Eastern Tibetan weather. *J. West China Border Res. Soc.* 145–150.

DURST, C. S. (1954). Variation of wind with time and distance. *Geophys. Mem. London* **12**, No. 93.

DURST, C. S., AND JOHNSON, D. H. (1959). The preparation of statistical wind forecasts and an assessment of their accuracy in comparison with forecasts made by synoptic techniques. *Meteorol. Off., London, Prof. Notes No.* 124.

EKMANN, V. W. (1905). On the influence of the earth's rotation on ocean currents. *Ark. Mat. Astron. Fys.* **2**, No. 11.

ELDRIDGE, R. H. (1957). A synoptic study of West African disturbance lines. *Quart. J. Roy. Meteorol. Soc.* **83**, 303–314.

ELIOT, J. (1895). A preliminary discussion of certain oscillatory changes of pressure of long period and of short period in India. *Mem. India Meteorol. Dept.* **6**, Part 2, 89–160.

ELIOT, J. (1905). A preliminary investigation of the more important features of the meteorology of southern Asia, the Indian Ocean and neighbouring countries during the period 1892–1902. *Mem. India Meteorol. Dept.* **16**, 185–307.

EMMONS, G., AND MONTGOMERY, R. B. (1947). Note on the physics of fog formation. *J. Meteorol.* **4**, 206.

EUWE, W. (1949). Forecasting rainfall in the periods December–January–February and April–May–June for parts of Celebes and South Borneo. *Kon. Magnet. Meteorol. Observ. Batavia, Verhand. No.* 38.

FALLS, R. (1970). Some synoptic models—North Australia. *In Proc. Symp. Tropical Meteorol. Honolulu*, pp. E V–1–8. Amer. Meteorol. Soc., Boston, Massachusetts.

FANKHAUSER, J. C. (1964). On the Motion and Predictability of Convective Systems as Related to the Upper Winds in a Case of Small Turning of Wind with Height. Nat. Severe Storms Project, Kansas City Rep. 21.

FINDLATER, J. (1968). The month to month variation of mean winds at low level over eastern Africa. *East African Meteorol. Dept. Tech. Mem. No.* 12.

FITZROY, R. (1863). "The Weather Book: A Manual of Practical Meteorology." Longman, Green, London.

FLOHN, H. (1960a). Recent investigations on the mechanism of the "summer monsoon" of southern and eastern Asia. *In* "Monsoons of the World" (S. Basu *et al.*, eds.), pp. 75–88. Hind Union Press, New Delhi.

FLOHN, H. (1960b). The structure of the intertropical convergence zone. *In* "Tropical Meteorology in Africa" (D. J. Bargman, ed.), pp. 244–252. Munitalp Foundation, Nairobi.

FLOHN, H. (1960c). Equatorial westerlies over Africa, their extension and significance. *In* "Tropical Meteorology in Africa" (D. J. Bargman, ed.), pp. 253–267. Munitalp Foundation, Nairobi.

FLOHN, H. (1964a). Zur Interpretation und raumlichen Verteilung statischer Parameter der Hohenwindverteilung. *Beitr. Phys. der Atmosphäre* **37**, 17–29.

FLOHN, H. (1964b). Über die Ursachen der aridität Nordost-Afrikas. *Würzburger Geogr. Arb.* **12**, 25–41.

FLOHN, H. (1964c). Investigations on the tropical easterly jet. *Bonner Meteorol. Abhand. No.* 4.

FLOHN, H. (1968). Contributions to a Meteorology of the Tibetan Highlands. Dept. Atmos. Sci., Colorado State Univ. Atmos. Sci. Paper No. 130.

FLOHN, H., AND STRÜNING, J.-O. (1969). Investigations on the atmospheric circulation above Africa. *Bonner Meteorol. Abhand. No.* 10.

Fox, T. (1969). An example of a medium level westerly wave over South and Central Africa. *Lusaka, Meteorol. Notes, Ser. A, No.* 2.

FREEMAN, M. H. (1961). A graphical method of objective forecasting derived by statistical techniques. *Quart. J. Roy. Meteorol. Soc.* **87**, 393–400.

FROLOW, S. (1951). Régimes bariques dans deux réseaux intertropicaux éloignés. *La Météorologie* **23**, 153–173.

FROLOW, S. (1960). La non existence de la mousson à Madagascar. *In* "Monsoons of the World" (S. Basu *et al.*, eds.), pp. 101–104. Hind Union Press, New Delhi.

FROST, R. (1953). Upper air circulation in low latitudes in relation to certain climatological discontinuities. *Meteorol. Off., London, Prof. Notes No.* 107.

FROST, R. (1969). Wind flow over tropical Africa. *Lusaka, Meteorol. Notes, Ser. A, No. 3.*

FRYER, J. (1698). "A New Account of East-India and Persia." Chiswell, London.

FUJITA, T., WATANABE, K., AND IZAWA, T. (1969). Formation and structure of equatorial anticyclones caused by large-scale cross-equatorial flows determined by ATS-1 photographs. *J. Appl. Meteorol.* **8**, 649–667.

FULKS, J. R. (1951). The instability line. *Compendium Meteorol.*, 647–652.

GARNIER, B. J. (1967). Weather conditions in Nigeria. *McGill Univ. Dept. Geography Climatolog. Res. Series No. 2.*

GARSTANG, M. (1965). Distribution and Mechanism of Energy Exchange between the Tropical Oceans and Atmosphere. Dept. Meteorol. Florida State Univ., Tallahassee.

GENÈVE, R. (1966). Problèmes de la météorologie tropicale. *La Météorologie* **8**, 47–54.

GENTRY, R. C. (1964a). On the Momentum and Energy Balance in Hurricane Helene. Nat. Hurricane Res. Lab. Rep. No. 69.

GENTRY, R. C. (1964b). Forecasting the movement of tropical cyclones. *In Proc. Symp. Tropical Meteorol.* (J. W. Hutchings, ed.), pp. 683–701. New Zealand Meteorol. Serv., Wellington.

GICHUIYA, S. N. (1970). Easterly disturbances in the southeast monsoon. *In Proc. Symp. Tropical Meteorol. Honolulu*, pp. E XIII–1–6. Amer. Meteorol. Soc., Boston, Massachusetts.

GILCHRIST, A. (1960). Contour charts for July 1960 in the West African area. *Tech. Notes Brit. W. African Meteorol. Serv.* **19**.

GLOVER, J., ROBINSON, P., AND HENDERSON, J. P. (1954). Provisional maps of the reliability of annual rainfall in East Africa. *Quart. J. Roy. Meteorol. Soc.* **80**, 602–609.

GODSKE, C. L., BERGERON, T., BJERKNES, J., AND BUNDGAARD, R. C. (1957). "Dynamic Meteorology and Weather Forecasting." Amer. Meteorol. Soc., Boston, Massachusetts.

GOLDSTEIN, S. (ed.) (1938). "Modern Developments in Fluid Mechanics." Clarendon Press, Oxford.

GORDON, A. H., AND TAYLOR, R. C. (1970). Numerical steady-state friction layer trajectories over the oceanic tropics as related to weather. *Int. Indian Ocean Expedition, Meteorol. Monogr.* **7**.

HALL, M. (1916). Notes on Hurricanes, Earthquakes, and Other Physical Occurrences in Jamaica up to the Commencement of the Weather Service, 1880, with Brief Notes in Continuation up to the End of 1915. Publ. Jamaica Meteorol. Serv. No. 455.

HAMILTON, R. A., AND ARCHBOLD, J. W. (1945). Meteorology of Nigeria and adjacent territory. *Quart. J. Roy. Meteorol. Soc.* **71**, 231–264.

HANEL, R., AND CONRATH, B. (1969). Interferometer experiment on Nimbus 3: preliminary results. *Science* **165**, 1258–1260.

HANN, J. (1908). "Handbuch der Klimatologie," Vol. 1. Engelhorn, Stuttgart.

HANSON, D. M. (1963). The Use of Meteorological Satellite Data in Analysis and Forecasting. U.S. Weather Bur., Off. of Forecast Develop., Tech. Note No. 13.

HARRIS, B. E., SADLER, J. C., BRETT, W. R., HO, F. P., AND ING. G. (1969). Role of the Synoptic Scale in Convection over Southeast Asia during the Summer Monsoon. Hawaii Inst. Geophys. Rep. 69–9.

HARRIS, B. E., ING, G., AND NERALLA, V. R. (1970). The influence of the synoptic scale on convection over Southeast Asia during the summer monsoon. *In Proc. Conf. Summer Monsoon Southeast Asia* (C. S. Ramage, ed.), pp. 255–270. Navy Weather Res. Facility, Norfolk, Virginia.

HART, T. J., AND CURRIE, R. I. (1960). The Benguela Current. *Discovery Rep.* **31**, 123–289.

HASTENRATH, S. L. (1968). On mean meridional circulation in the tropics. *J. Atmos. Sci.* **25**, 979–983.

HAWKINS, H. F., AND RUBSAM, D. T. (1968). Hurricane Hilda, 1964. 1. Genesis, as revealed by satellite photographs, conventional and aircraft data. *Mon. Weather Rev.* **96**, 428–452.

HELLAND-HANSEN, B., AND NANSEN, F. (1920). Temperature variation in the North Atlantic Ocean and in the atmosphere, introductory studies on the causes of climatological variations. *Smithsonian Inst. Misc. Collect.* **70**, No. 4 Publ. 2537.

HENDERSON, J. P. (1949). Some aspects of climate in Uganda with special reference to rainfall. *East African Meteorol. Dept. Mem. No. 5.*

HENDL, M. (1963). "Systematische Klimatologie." Veb Deutscher Verlag der Wissenschaften, Berlin.

HENRY, W. K., AND GRIFFITHS, J. F. (1963). Research on Tropical Rainfall Patterns and Associated Meso-Scale Systems. Rep. No. 4 to U. S. Army Electron. Res. and Develop. Lab., Texas A & M Res. Foundation, College Station, Texas.

HEYWOOD, G. S. P. (1953). Surface pressure patterns and weather around the year in Hong Kong. *Roy. Observ. Hong Kong. Tech. Mem. No. 6.*

HILL, H. W. (1964). The weather in lower latitudes of the Southwest Pacific associated with the passage of disturbances in the middle latitude westerlies. *In Proc. Symp. Tropical Meteorol.* (J. W. Hutchings, ed.), pp. 352–365. New Zealand Meteorol. Serv., Wellington.

HO, F. P., AND MURAKAMI, T. (1970). Interaction between middle-latitude disturbances and tropical circulation. *In Proc. Symp. Tropical Meteorol. Honolulu*, pp. NI–1–6. Amer. Meteorol. Soc., Boston, Massachusetts.

HOFMEYR, W. L. (1961). Statistical analyses of upper air temperatures and winds over tropical and subtropical Africa. *Notos* **10**, 123–149.

HSIEH, Y.-P., (1949). An investigation of a selected cold vortex over North America. *J. Meteorol.* **6**, 401–410.

HSU, C. (1965). Mei-Yü of the middle and lower Yangtze Valley in the past 80 years (in Chinese). *Acta Meteorol. Sinica* **35**, 507–518.

HUKE, R. E. (1965). Rainfall in Burma. *Geogr. Publ. Dartmouth* **2**.

India Meteorological Department (1953). "Climatological Tables of Observatories in India." Manager of Publications, Delhi.

India Meteorological Department (1962). Monthly and annual normals of rainfall and of rainy days. *Mem. India Meteorol. Dept.* **31**, Part 3.

IYER, V. D. (1931). Rainfall of Siam. *India Meteorol. Dept. Sci. Notes* **4**, 69–85.

IYER, V. D., AND DASS, I. (1946). Diurnal variation of rainfall at Mahabeleshwar. *India Meteorol. Dept. Sci. Notes* **9**, 37–42.

IYER, V. D., AND ZAFAR, M. (1940). Distribution of heavy rainfall over India. *India Meteorol. Dept. Sci. Notes* **7**, 109–118.

IYER, V. D., AND ZAFAR, M. (1946). Diurnal variation of rainfall at Simla. *India Meteorol. Dept. Sci. Notes* **9**, 33–36.

JAGANNATHAN, P., AND KHANDEKAR, M. L. (1962). Predisposition of the upper air structure in March to May over India to the subsequent monsoon rainfall of the peninsula. *Indian J. Meteorol. Geophys.* **13**, 305–316.

JALU, R. (1965). Note sur le déclenchement des dépressions tropicales sahariennes. *La Météorologie* **7**, 113–128.

JALU, R., AND DETTWILLER, J. (1965). Advection froide vers les basses latitudes (Mars 1963). *La Météorologie* **6**, 159–167.

Japan Meteorological Agency (1963). Mean Maps for January 1963. Tokyo.

Japan Meteorological Agency (1964). Mean Maps for January 1964. Tokyo.

Japan Meteorological Agency (1967). Mean Maps for March 1967. Tokyo.

JEANDIDIER, G., AND RAINTEAU, P. (1957). Prévision du temps sur le bassin du Congo. *France, Météorol. Nat., Monogr.* **9**.

JOHNSON, D. H. (1962). Rain in East Africa. *Quart. J. Roy. Meteorol. Soc.* **88**, 1–19.

JOHNSON, D. H. (1964a). Weather systems of West and central Africa. *In Proc. Symp. Tropical Meteorol.* (J. W. Hutchings, ed.), pp. 339–346. New Zealand Meteorol. Serv., Wellington.

JOHNSON, D. H. (1964b). Forecasting weather in East Africa. *In* High-level forecasting. for turbine-engined aircraft operations over Africa and the Middle East, Vol. 1, pp. 83–94. *W.M.O. Tech. Note No.* 64.

JOHNSON, D. H. (1964c). Forecasting weather in West Africa. *In* High-level forecasting for turbine-engined aircraft operations over Africa and the Middle East, Vol. I. pp. 95–103. *W.M.O. Tech. Note No.* 64.

JOHNSON, D. H. (1964d). Commentary on the analysis work relating to tropical Africa. *In* High-level forecasting for turbine-engined aircraft operations over Africa and the Middle East, Vol. 2, pp. 21–31. *W.M.O. Tech. Note No.* 64.

JOHNSON, D. H., AND MÖRTH, H. T. (1960). Forecasting research in East Africa. *In* " Tropical Meteorology in Africa " (D. J. Bargman, ed.), pp. 56–137. Munitalp Foundation, Nairobi.

KAO, Y.-H. (1951). Forecast of typhoon movements on the basis of statistics (in Chinese). *Chi-hsiang Hsueh-pao* **21**.

KAO, Y.-H. (1958). On the clear weather over eastern Asia during high autumn (in Chinese). *Acta Meteorol. Sinica* **29**, 83–92.

KAO, Y.-H., AND KUO, C.-Y. (1958). Autumn rain areas in China (in Chinese). *Acta Meteorol. Sinica* **29**, 264–273.

KAO, Y.-H., *et al.* (1962). Some problems on the monsoons of East Asia. *Collect. Papers Inst. Geophys. Meteorol., Acad. Sinica* **5**. [Engl. Transl. Emm-66-124, Oriental Science Library, Emmanuel College, Boston.]

KASAHARA, A. (1961). A numerical experiment on the development of a tropical cyclone. *J. Meteorol.* **18**, 259–282.

KEEGAN, T. J. (1958). Arctic synoptic activity in winter. *J. Meteorol.* **15**, 513–521.

KELLY, W. L. (1964). The double jet: a proposed synoptic model for the tropics. *In Proc. Symp. Tropical Meteorol.* (J. W. Hutchings, ed.), p. 417. New Zealand Meteorol. Serv., Wellington.

KENDREW, W. G. (1961). "The Climate of the Continents," 5th ed. Clarendon Press, Oxford.

KHROMOV, S. P. (1957). Die geographische Verbreitung der Monsune. *Petermanns Geogr. Mitt.* **101**, 234–237.

KLEIN, W. H. (1957). Principal Tracks and Frequencies of Cyclones and Anticyclones in the Northern Hemisphere. U.S. Weather Bur., Res. Pap. No. 40.

KOO, C.-C., CHEN, Y.-S., AND SHU, Y.-F. (1958). A frontal θ_{SE} chart and its application to the analysis of the upper boundary of cold-wave fronts in China (in Chinese). *Acta Meteorol. Sinica* **29**, 44–56.

KOTESWARAM, P. (1950). Upper level "lows" in the Indian area during SW monsoon season and " breaks " in the monsoon. *Indian J. Meteorol. Geophys.* **1**, 162–164.

KOTESWARAM, P. (1958). The easterly jet stream in the tropics. *Tellus* **10**, 43–57.

KOTESWARAM, P., AND BHASKARA RAO, N. S. (1963). Formation and structure of Indian summer monsoon depressions. *Aust. Meteorol. Mag.* **41**, 62–75.

KOTESWARAM, P., AND GEORGE, C. A. (1958). On the formation of monsoon depressions in the Bay of Bengal. *Indian J. Meteorol. Geophys.* **9**, 9–22.

KOTESWARAM, P., AND GEORGE, C. A. (1960). A case study of a monsoon depression in the Bay of Bengal. *In* " Monsoons of the World " (S. Basu *et al.*, eds.), pp. 145–156. Hind Union Press, New Delhi.

KREITZBERG, C. W. (1970). Mesoscale and diurnal variations deduced from the Saigon area rawinsonde data. *In Proc. Conf. Summer Monsoon Southeast Asia* (C. S. Ramage, ed.), pp. 271–282. Navy Weather Res. Facility, Norfolk, Virginia.

KRISHNAMURTI, T. N. (1969). An experiment in numerical prediction in equatorial latitudes. *Quart. J. Roy. Meteorol. Soc.* **95**, 594–620.

KUO, H. L. (1965). On formation and intensification of tropical cyclones through latent heat release by cumulus convection. *J. Atmos. Sci.* **22**, 40–63.

KURASHIMA, A. (1968). Studies on the winter and summer monsoons in East Asia based on dynamic concept. *Geophys. Mag.* **34**, 145–235.

LALLY, V. E., LITCHFIELD, E. W., AND SOLOT, S. B. (1966). The Southern Hemisphere GHOST experiment. *W.M.O. Bull.* **15**, 124–128.

LA SEUR, N. E. (1967). Equivalent potential temperature as a measure of convective and synoptic scale disturbances of the tropical atmosphere. *In Proc. 1967 Army Conf. Tropical Meteorol.* (H. W. Hiser and H. P. Gerrish, eds.), pp. 196–201. Inst. of Marine Sci., Univ. of Miami, Miami, Florida.

LAVOIE, R. L., AND WIEDERANDERS, C. J. (1960). Objective Wind Forecasting over the Tropical Pacific. Hawaii Inst. Geophys. Rep. No. 12.

LEBEDEV, A. N., AND SOROCHAN, O. G., eds. (1967). "Klimaty Afriki." Gidrometeoizdat, Leningrad.

LEE, C., AND WAN, M. W. (1955). The climatic types of floods and droughts in the valleys of the Yangtze and the Huai Rivers (in Chinese). *Acta Geogr. Sinica* **21**, 245–258.

LEE, J. (1932). Die grosse Uberschwemmung des Jangtsegebietes im Juli 1931. *Meteorol. Z.* **49**, 234–237.

LEESE, J. A., BOOTH, A. L., AND GODSHALL, F. A. (1969). Archiving and Climatological Applications of Meteorological Satellite Data. Environmental Sci. Serv. Administration, Rockville, Maryland.

LEIPPER, D. F. (1967). Observed ocean conditions in hurricane Hilda, 1964. *J. Atmos. Sci.* **24**, 182–196.

LI, G.-w. (1965). A study of the thermal low over Southwest China and related forecasting problems (in Chinese). *Acta Meteorol. Sinica* **35**, 126–131.

LING, S. H. (1965). Diurnal variation of upper air temperature, pressure and humidity over Taipei. *Nat. Taiwan Univ. Dept. Geogr. Meteorol. Sci. Rep.* **3**, 58–72.

LORENZ, E. N. (1955). Available potential energy and the maintenance of the general circulation. *Tellus* **7**, 157–167.

MAYENÇON, R. (1961). Conditions synoptiques donnent lieu à des précipitations torrentielles au Sahara. *La Météorologie,* **4**, 171–180.

McCABE, J. T. (1961). (1) A systematic approach to the use of climatology and persistence in terminal weather forecasting. (2) An application of the climatology of typhoon movement to operational forecasting and estimation of typhoon threat to a particular installation. *In U.S.-Asian Weather Symp.* First Weather Wing, USAF, San Francisco, California.

McCUTCHAN, M. H., AND HELFMAN, R. S. (1969). Synoptic-Scale Weather Disturbances that Influence Fire Climate in Southeast Asia During Normally Dry Periods—A Preliminary Report. Pacific Southwest Forest and Range Exp. Station, Berkeley, California.

McDONALD, W. F. (1938). Atlas of Climatic Charts of the Oceans. U.S. Weather Bur., Washington, D.C.

McQUAIN, R. H. (1967). ATS-1 camera experiment successful. *Bull. Amer. Meteorol. Soc.* **48**, 74–79.

MEINARDUS, W. (1893). Beiträge zur kenntniss der Klimatischen Verhältnisse des Nordöstlichen Teils des Indischen Ozeans. *Arch. Deut. Seewarte* **16**, No. 7.

Meteorological Office (1947). Monthly Meteorological Charts of the Western Pacific. H.M. Stationery Office, London.

Meteorological Office (1949). Monthly Meteorological Charts of the Indian Ocean. H.M. Stationery Office, London.

MILLER, B. I. (1958). The Three-Dimensional Wind Structure around a Tropical Cyclone. Nat. Hurricane Res. Project Rep. No. 15.

MILLER, B. I.(1964). A study of the filling of hurricane Donna (1960) over land. *Mon. Weather Rev.* **92**, 389–406.

MILLER, B. I. (1967). Characteristics of hurricanes. *Science* **157**, 1389–1399.

MILLER, F. R., AND KESHAVAMURTHY, R. N. (1968). Structure of an Arabian Sea summer monsoon system. *Int. Indian Ocean Expedition, Meteorol. Monogr.* **1**.

MILLER, R. G. (1958). Statistics and predictability of weather. *In* "Studies in Statistical Weather Prediction," pp.137–153. Travelers Weather Res. Center, Hartford, Connecticut.

MINTZ, Y. (1968). Very long-term global integration of the primitive equations of atmospheric motion: an experiment in climatic simulation. *Meteorol. Monogr.* **8**, 20–36.

MIYAKODA, K., SMAGORINSKY, J., STRICKLER, R. F., AND HEMBREE, G. D. (1969). Experimental extended predictions with a nine-level hemispheric model. *Mon. Weather Rev.* **97**, 1–76.

MONTRIWATE, T. (1963). Report of severe tropical storm Harriet, 1962. *In Final Rep. Third U.S.-Asian Weather Symp.* (N. M. Burgner and W. E. Warner, eds.). First Weather Wing, USAF, San Francisco, California.

MOOLEY, D. A. (1957), The role of western disturbances in the prediction of weather over India during the different seasons. *Indian J. Meteorol. Geophys.* **8**, 253–260.

MUFFATTI, A. H. J. (1964). Aspects of the subtropical jet stream over Australia. *In Proc. Symp. Tropical Meteorol.* (J. W. Hutchings, ed.), pp. 72–88. New Zealand Meteorol. Serv., Wellington.

MULL, S. AND RAO, Y. P. (1949). Indian tropical storms and zones of heavy rainfall. *Indian J. Phys.* **23**, 371–377.

MURAKAMI, T. (1958). The sudden change of upper westerlies near the Tibetan plateau at the beginning of summer season. *J. Meteorol. Soc. Japan* **36**, 239–247.

MURAKAMI, T. (1959a). The energy budget over the Far East during the rainy season. *J. Meteorol. Soc. Japan (Ser. 2)* **37**, 83–95.

MURAKAMI, T. (1959b). The general circulation and water vapor balance over the Far East during the rainy season. *Geophys. Mag.* **29**, 131–171.

MURAKAMI, T. (1969). Initial Adjustment of Data in the Tropics. Hawaii Inst. Geophys. Rep. 69–23.

MURAKAMI, T., GODBOLE, R. V., AND KELKAR, R. R. (1970). Numerical simulation of the monsoon along 80E. *In Proc. Conf. Summer Monsoon Southeast Asia* (C. S. Ramage, ed.), pp. 39–51. Navy Weather Res. Facility, Norfolk, Virginia.

NAKAMURA, I. (1967). Minimum Temperature Forecasting at Naha, Okinawa by Statistical Techniques. M.S. Thesis, Dept. of Geosciences, Univ. of Hawaii, Honolulu.

NAMIAS, J. (1963). Large-scale air–sea interactions over the North Pacific from summer 1962 through the subsequent winter. *J. Geophys. Res.* **68**, 6171–6186.

NAMIAS, J. (1968). Long-range weather forecasting—history, current status and outlook. *Bull. Amer. Meteorol. Soc.* **49**, 438–470.

NAMIAS, J. (1969). Seasonal interactions between the North Pacific Ocean and the atmosphere during the 1960's. *Mon. Weather Rev.* **97**, 173–192.

NAMIAS, J., AND CLAPP, P. F. (1949). Confluence theory of the high tropospheric jet stream. *J. Meteorol.* **6**, 330–336.

NARASIMHAM, V. L., AND ZAFAR, M. (1947). An analysis of the hourly rainfall records at Poona. *India Meteorol. Dept. Sci. Notes* **9**, 94–110.

National Aeronautics and Space Administration (1966). Significant Achievements in Satellite Meteorology, 1958–1964. Washington, D.C.

Navy Weather Research Facility Staff (1969). The Diagnosis and Prediction of SE Asia Northeast Monsoon Weather. Norfolk, Virginia.

NEHRU, J. (1942). "The Unity of India; Collected Writings 1937–1940." Lindsay Drummond, London.

NEUMANN, J. (1951). Land breezes and nocturnal thunderstorms. *J. Meteorol.* **8**, 60–67.

NEVIÈRE, E. (1959). Note sur la variation diurne du front intertropical en Afrique équatoriale française. Relations entre les courants rapides et les types de temps en Afrique équatoriale. *France, Météorol. Nat. Monogr.* **14**.

NEWTON, C. W. (1950). Note on the mechanism of nor-westers of Bengal. *Indian J. Meteorol. Geophys.* **2**, 48–50.

NEWTON, C. W. (1960). Morphology of thunderstorms and hailstorms as affected by vertical wind shear. *Amer. Geophys. Union Geophys. Monogr.* **5**, 339–347.

NEWTON, C. W. (1967). Severe convective storms. *Advan. Geophys.* **12**, 257–308.

NEWTON, C. W., AND KATZ, S. (1958). Movement of large convective rainstorms in relation to winds aloft. *Bull. Amer. Meteorol. Soc.* **39**, 129–136.

NEWTON, C. W., AND NEWTON, H. R. (1959). Dynamical interactions between large convective clouds and environment with vertical shear. *J. Meteorol.* **16**, 483–496.

NICHOLSON, J. R. (1970). Diurnal variation of winds over India. *Int. Indian Ocean Expedition, Meteorol. Monogr.* **9** (in preparation).

NIEUWOLT, S., POOI, L. Y., AND TEE, S. C. (1967). Rainfall in East Malaysia and Brunei. *Meteorol. Serv. Singapore Mem. No.* 8.

NORMAND, C. W. B. (1953). Monsoon seasonal forecasting. *Quart. J. Roy. Meteorol. Soc.* **79**, 463–473.

O'CONNOR, J. F. (1963). The weather and circulation of January 1963. *Mon. Weather Rev.* **91**, 209–217.

OKADA, T. (1910). On the Bai-U or rainy season in Japan *Bull. Central Meteorol. Obs. Japan No.* 5.

OOYAMA, K. (1969). Numerical simulation of the life cycle of tropical cyclones. *J. Atmos. Sci.* **26**, 3–40.

OTANI, T. (1954). Converging line of the northeast trade wind and converging belt of the tropical air current. *Geophys. Mag.* **25**, 1–122.

PALMÉN, E. (1949). Origin and structure of the high-level cyclones south of the maximum westerlies. *Tellus* **1**, 22–31.

PALMÉN, E. (1956). Formation and development of tropical cyclones. *In Proc. Tropical Cyclone Symp., Brisbane*, pp. 213–231. Bur. Meteorol., Melbourne, Australia.

PALMÉN, E., AND NAGLER, K. M. (1949). The formation and structure of a large-scale disturbance in the westerlies. *J. Meteorol.* **6**, 227–242.

PALMÉN, E., AND NEWTON, C. W. (1969). "Atmospheric Circulation Systems, their Structure and Physical Interpretation." Academic Press, New York.

PALMER, C. E. (1951). On high-level cyclogenesis originating in the tropics. *Trans. Amer. Geophys. Union* **32**, 683–696.

PALMER, C. E., AND OLMSTEDE, W. D. (1956). The simultaneous oscillation of barometers along and near the equator. *Tellus* **4**, 495–507.

PALMER, C. E., WISE, C. W., STEMPSON, L. J., AND DUNCAN, G. H. (1955). The Practical Aspect of Tropical Meteorology. U.S. Air Force Surv. Geophys. No. 76.

PANOFSKY, H. (1957). "Introduction to Dynamic Meteorology." Penn. State University Press, University Park, Pennsylvania.

PARTHASARATHY, K. (1960). Some aspects of the rainfall in India during the southwest monsoon. *In* "Monsoons of the World" (S. Basu *et al.*, eds.), pp. 185–194. Hind Union Press, New Delhi.

PÉDELABORDE, P. (1958). "Les Moussons." Librairie Armand Colin, Paris.

PEDGLEY, D. E. (1969a). Diurnal variation of the incidence of monsoon rainfall over the Sudan. *Meteorol. Mag.* **98**, 97–107, 129–134.

PEDGLEY, D. E. (1969b). Cyclones along the Arabian coast. *Weather* **24**, 456–468.

PEDGLEY, D. E. (1970). An unusual monsoon disturbance over southern Arabia. *In Proc. Symp. Tropical Meteorol. Honolulu*, pp. E VI–1–6. Amer. Meteorol. Soc., Boston, Massachusetts.

PETTERSSEN, S. (1939). Some aspects of formation and dissipation of fog. *Geofys. Publikasjoner* **12**, No. 10.

PETTERSSEN, S. (1956). "Weather Analysis and Forecasting." McGraw-Hill, New York.

PIKE, A. C. (1968). A Numerical Study of Tropical Atmospheric Circulations. Univ. Miami, Inst. Atmos. Sci., Sci. Rep. No. 1 for USAF Cambridge Res. Lab. Contr. No. F19628-68-C-0144.

PIKE, A. C. (1970). The intertropical convergence zone studied with an interacting atmosphere and ocean model. *In Proc. Symp. Tropical Meteorol. Honolulu*, pp. F III–1–5. Amer. Meteorol. Soc., Boston, Massachusetts.

PISHAROTY, P. R., AND ASNANI, G. C. (1957). Rainfall around monsoon depressions over India. *Indian J. Meteorol. Geophys.* **8**, 15–20.

PISHAROTY, P. R., AND DESAI, B. N. (1956). "Western disturbances" and Indian weather. *Indian J. Meteorol. Geophys.* **7**, 331–338.

POINTER, J. (1738). "A Rational Account of the Weather." Aaron Ward, London.

PORTIG, W. H. (1963). Thunderstorm frequency and amount of precipitation in the tropics, especially in the African and Indian monsoon regions. *Arch. Meteorol. Geophys. Bioklimatol. B* **13**, 21–35.

POWELL, J., AND PEDGLEY, D. E. (1969). A year's weather at Termit, Republic of Niger. *Weather* **24**, 247–254.

PREEDY, B. H. (1966). Far East Air Force Climatology. Headquarters, Far East Air Force, Singapore.

PYLE, R. L. (1965). Meteorological satellite data, archiving and availability. *Bull. Amer. Meteorol. Soc.* **46**, 707–713.

RAGHAVAN, K. (1967). A climatological study of severe cold waves in India. *Indian J. Meteorol. Geophys.* **18**, 91–96.

RAINBIRD, A. F. (1968). Weather Disturbances over Tropical Continents and Their Effect on Ground Conditions. Dept. Atmos. Sci., Colorado State Univ., Fort Collins.

RAINEY, R. C. (1963). Meteorology and the migration of desert locusts. *W.M.O. Tech. Note No.* 54.

RAI SIRCAR, N. C. (1956). A note on the vertical structure of a few disturbances of the Bay of Bengal. *Indian J. Meteorol. Geophys.* **7**, 37–42.

RAM, N. (1929). Frequency of thunderstorms in India. *India Meteorol. Dept. Sci. Notes* **1**, 49–55.

RAMAGE, C. S. (1951). Analysis and forecasting of summer weather over and in the neighborhood of South China. *J. Meteorol.* **8**, 289–299.

RAMAGE, C. S. (1952a). Diurnal variation of summer rainfall over East China, Korea and Japan. *J. Meteorol.* **9**, 83–86.

RAMAGE, C. S. (1952b). Variation of rainfall over South China through the wet season. *Bull. Amer. Meteorol. Soc.* **33**, 308–311.

RAMAGE, C. S. (1952c). Relationship of general circulation to normal weather over southern Asia and the western Pacific during the cool season. *J. Meteorol.* **9**, 403–408.

RAMAGE, C. S. (1954). Non-frontal crachin and the cool season clouds of the China Seas. *Bull. Amer. Meteorol. Soc.* **35**, 404–411.

RAMAGE, C. S. (1955). The cool-season tropical disturbances of Southeast Asia. *J. Meteorol.* **12**, 252–262.

RAMAGE, C. S. (1956). Review of methods of tropical synoptic analysis (with particular reference to tropical cyclones). *In Proc. Tropical Cyclone Symp., Brisbane,* pp. 75–96. Bur. Meteorol., Melbourne, Australia.

RAMAGE, C. S. (1959). Hurricane development. *J. Meteorol.* **16**, 227–237.

RAMAGE, C. S. (1962). The subtropical cyclone. *J. Geophys. Res.* **67**, 1401–1411.

RAMAGE, C. S. (1964a). Diurnal variation of summer rainfall of Malaya. *J. Trop. Geogr.* **19**, 62–68.

RAMAGE, C. S. (1964b). Relations between weather, winds and pressures in low latitudes. *Nature* **201**, 1206–1207.

RAMAGE, C. S. (1964c). Some preliminary research results from the International Meteorological Center. *In Proc. Symp. Tropical Meteorol.* (J. W. Hutchings, ed.), pp. 403–408. New Zealand Meteorol. Serv., Wellington.

RAMAGE, C. S. (1966). The summer atmospheric circulation over the Arabian Sea. *J. Atmos. Sci.* **23**, 144–150.

RAMAGE, C. S. (1968a). Problems of a monsoon ocean. *Weather* **23**, 28–37.

RAMAGE, C. S. (1968b). Role of a tropical "Maritime Continent" in the atmospheric circulation. *Mon. Weather Rev.* **96**, 365–370.

RAMAGE, C. S. (1969a). Summer drought over western India. *In* "Eclectic Climatology" (A. Court, ed.), pp. 41–54. Oregon State Univ. Press, Corvallis, Oregon.

RAMAGE, C. S. (1969b). Indian Ocean surface meteorology. *Oceanogr. Mar. Biol. Ann. Rev.* **7**, 11–30.

RAMAGE, C. S. (1970). Meteorology of the South Pacific. *In* "Scientific Exploration of the South Pacific," pp. 16–29. Nat. Acad. Sci., Washington, D.C.

RAMAGE, C. S., AND RAMAN, C. R. V. (1970). "International Indian Ocean Expedition, Meteorological Atlas," Vol. 2, "Upper Air." National Science Foundation, Washington, D.C.

RAMAMURTHI, K. M., AND JAMBUNATHAN, R. (1967). On the onset of the southwest monsoon rains along extreme southwest coast of peninsular India. *In Proc. Symp. Meteorol. Results Int. Indian Ocean Expedition* (P. R. Pisharoty, ed.), pp. 374–379. India Meteorol. Dept., New Delhi.

RAMAMURTHI, K. M., KESHAVAMURTHY, R. N., AND JAMBUNATHAN, R. (1967). Some distinguishing features of strong and weak monsoon regimes over India and neighbourhood. *In Proc. Symp. Meteorol. Results Int. Indian Ocean Expedition* (P. R. Pisharoty, ed.), pp. 350–361. India Meteorol. Dept., New Delhi.

RAMAN, C. R. V. (1970). Structure of the summer trough system over the northern Indian Ocean. *Int. Indian Ocean Expedition, Meteorol. Monogr.* **10** (in preparation).

RAMAN, C. R. V., AND RAMANATHAN, Y. (1964). Interaction between lower and upper tropical tropospheres. *Nature* **204**, 31–35.

RAMANATHAN, K. R., AND RAMAKRISHNAN, K. P. (1932). The Indian southwest monsoon and the structure of depressions associated with it. *Mem. India Meteorol. Dept.* **26**, 13–36.

RAMANATHAN, K. R., AND VENKITESHWARAN, S. P. (1948). Climatological Charts of the Indian Monsoon Area. India Meteorol. Dept., New Delhi.

RAMASWAMY, C. (1958). A preliminary study of the behavior of the Indian southwest monsoon in relation to the westerly jet stream. *Geophysica* **6**, 455–476.

RAMASWAMY, C. (1962). Breaks in the Indian summer monsoon as a phenomenon of interaction between the easterly and the subtropical westerly jet streams. *Tellus* **14**, 337–349.

RAMASWAMY, C. (1966). The problem of fronts in the Indian atmosphere. *Indian J. Meteorol. Geophys.* **17**, 151–170.

RAMASWAMY, C. (1967). On synoptic methods of forecasting the vagaries of southwest monsoon over India and neighbouring countries. *In Proc. Symp. Meteorol. Results Int. Indian Ocean Expedition* (P. R. Pisharoty, ed.), pp. 317–336. India Meteorol. Dept., New Delhi.

RAMASWAMY, C. (1969). The Problem of the Indian Southwest Monsoon. Indian Geophys. Union, Hyderabad, India.

RAMASWAMY, C., AND SURYANARAYANA, N. (1950). Rainfall at Peshawar. *Mem. India Meteorol. Dept.* **28**, 121–137.

RAMDAS, L. A. (1949). Rainfall of India: a brief review. *Indian J. Agr. Sci.* **29**, 1–19.

RANGANATHAN, C., AND SOUNDARAJAN, K. (1965). A study of a typical case of interaction of an easterly wave with a westerly trough during the postmonsoon period. *Indian J. Meteorol. Geophys.* **16**, 607–616.

RAO, K. N. (1965). Seasonal forecasting—India. WMO-IUGG Symposium on research and development aspects of long-range forecasting, pp. 17–30. *W.M.O. Tech. Note No.* 66.

RAO, K. N., AND RAMAN, P. K. (1959). Diurnal variation of rainfall in India. *In* " Meteorological and Hydrological Aspects of Floods and Droughts in India " (S. Basu, ed.), pp. 186–191. Manager of Publications, New Delhi.

RAO, Y. P., AND RAMAMURTI, K. S. (1968). " Forecasting Manual. Part 1: Climatology of India and Neighbourhood. 2: Climate of India." India Meteorol. Dept., Poona, India.

REED, R. J., AND KUNKEL, B. A. (1960). The Arctic circulation in summer. *J. Meteorol.* **17**, 489–506.

REFSDAL, A. (1930). Der Feuchtlabile Niederschlag. *Geofys. Publikasjoner* **5**, No. 12.

REID, J. L. (1967). Upwelling. *In* " International Dictionary of Geophysics " (S. K. Runcorn, ed.), pp. 1638–1640. Macmillan (Pergamon), New York.

REITER, E. R. (1959). Das ende des Indischen Sommermonsuns 1954 mit Daten der Osterreichischen Cho-Oyu-Expedition. *Ber. Deut. Wetterdienstes* **54**, 293–297.

REITER, E. R. (1963). " Jet-Stream Meteorology." Univ. of Chicago Press, Chicago, Illinois.

Rex, D. F. (1958). Vertical atmospheric motions in the equatorial central Pacific. *Geophysica* **6**, 479–501.

RIEHL, H. (1954). " Tropical Meteorology." McGraw-Hill, New York.

RIEHL, H., AND MALKUS, J. S. (1961). On the heat balance in the equatorial trough zone. *Geophysica* **6**, 503–538.

RIEHL, H., AND PEARCE, R. P. (1968). Studies on Interaction between Synoptic and Mesoscale Weather Elements in the Tropics. Colorado State Univ. Atmos. Sci. Paper No. 126.

RIEHL, H., AND SHAFER, R. J. (1944). The recurvature of tropical storms. *J. Meteorol.* **1**, 42–54.

RIEHL, H., *et al.* (1952). Forecasting in middle latitudes. *Meteorol. Monogr.* **1**, No. 5.

ROBINSON, G. D. (1967). Some current projects for global meteorological observation and experiment. *Quart. J. Roy. Meteorol. Soc.* **93**, 409–418.

RODEN, G. I. (1959). On the heat and salt balance of the California Current region. *J. Mar. Res.* **18**, 36–61.

ROLL, H. U. (1965). " Physics of the Marine Atmosphere." Academic Press, New York.

ROSENTHAL, S. L. (1964). Some attempts to simulate the development of tropical cyclones by numerical methods. *Mon. Weather Rev.* **92**, 1–21.

Royal Observatory, HONG KONG (1961). " Meteorological Results, 1957—Part 2—Upper Air Observations." Government Printer, Hong Kong.

RUBIN, M. J. (1955). An investigation of relationships between Northern and Southern Hemisphere parameters. *Notos* **4**, 122–126.

RUSH, R. E. (1967). An Objective Surface/1000 mb Analysis Designed to Retain Sub-Mesh Length Perturbations. M. S. Thesis, Dept. of Geosciences, Univ. of Hawaii, Honolulu.

RUSSELL, H. C. (1893). Moving anticyclones in the Southern Hemisphere. *Quart. J. Roy. Meteorol. Soc.* **19**, 23–34.

SADLER, J. C. (1962). Utilization of Meteorological Satellite Cloud Data in Tropical Meteorology. USAF Cambridge Res. Lab., Res. Note AFCRL, 62–829.

SADLER, J. C. (1963). Tiros Observations of the Summer Circulation and Weather Patterns of the Eastern North Pacific. Hawaii Inst. Geophys. Rept. No. 40.

SADLER, J. C. (1967). The Tropical Upper Tropospheric Trough as a Secondary Source of Typhoons and a Primary Source of Tradewind Disturbances. Hawaii Inst. Geophys. Rep. No. 67–12.

SADLER, J. C. (1969). Average cloudiness in the tropics from satellite observations. *Int Indian Ocean Expedition, Meteorol. Monogr.* **2**.

SADLER, J. C., BRETT, W. R., HARRIS, B. E., AND HO, F. P. (1968). Forecasting Minimum Cloudiness over the Red River During the Summer Monsoon. Hawaii Inst. Geophys. Rep. No. 68–16.

SAHA, K. (1968). On the instantaneous distribution of vertical velocity in the monsoon field and structure of the monsoon circulation. *Tellus* **20**, 601–620.

SANDSTRÖM, J. W. (1909). Über die Bewegung der Flüssigkeiten. *Ann. Hydrograph. Maritimen Meteorol.* **37**, 242–254.

SAWYER, J. S. (1947). Notes on the theory of tropical cyclones. *Quart. J. Roy. Meteorol. Soc.* **73**, 101–126.

SCHERHAG, R. (1948). " Neue Methoden der Wetteranalyse und Wetterprognose." Springer, Berlin.

SCHICK, M. (1953). Die Geographische Verbreitung des Monsuns. *Nova Acta Leopoldina* **16**, No. 12.

CHMIDT, F. H. (1949). On the theory of small disturbances in equatorial regions. *J. Meteorol.* **6**, 427–428.

Serviço Meteorológico de Angola (1955). "O Clima de Angola." Luanda, Angola.

SHAPAEV, V. M. (1960). Monsoon characteristics of the atmospheric circulation in the Soviet Arctic (in Russian). *Geograficheskoe Obshchestvo, SSR, Izv.* **92**, 176–180.

SHAW, W. N. (1936). "Manual of Meteorology," Vol. 2. "Comparative Meteorology," Cambridge Univ. Press, London and New York.

SIMPSON, G. C. (1921). The south-west monsoon. *Quart. J. Roy. Meteorol. Soc.* **47**, 151–172.

SIMPSON, J., GARSTANG, M., ZIPSER, E. J., AND DEAN, G. A. (1967). A study of a non-deepening tropical disturbance. *J. Appl. Meteorol.* **6**, 237–254.

SIMPSON, R. H. (1952). Evolution of the Kona storm, a subtropical cyclone. *J. Meteorol.* **9**, 24–35.

SOMERVELL, W. L., and ADLER, R. F. (1970). A Preliminary Survey of Southeast Asia Spring Transformation Season Weather. Navy Weather Res. Facility, Norfolk, Virginia.

SOURBEER, R. H., AND GENTRY, R. C. (1961). Rainstorm in southern Florida, January 21, 1957. *Mon. Weather Rev.* **89**, 9–16.

SOUTHERN, R. L. (1969). Personal communication.

SOUTHERN, R. L., KININMONTH, W. R., AND PESCOD, N. R. (1970). Derivation of convective forecasting models for northern Australia from a climatology of lightning discharges. *In Proc. Conf. Summer Monsoon Southeast Asia* (C. S. Ramage, ed.), pp. 239–254. Navy Weather Res. Facility, Norfolk, Virginia.

SREENIVASAIAH, B. N., AND SUR, N. K. (1939). A study of the duststorms of Agra. *Mem. India Meteorol. Dept.* **27**, 1–30.

Staff Members, Institute of Geophysics and Meteorology (1957–58). On the general circulation over eastern Asia. *Tellus* **9**, 432–446; **10**, 58–75, 299–312.

STOMMEL, H., AND WOOSTER, W. S. (1965). Reconnaissance of the Somali Current during the southwest monsoon. *Proc. Nat. Acad. Sci.* **54**, 8–13.

SUCKSTORFF, G. A. (1939). Die Ergebnisse der Untersuchungen an Tropischen Gewittern und einigen anderen Erscheinungen. *Gerlands Beitr. Geophys.* **55**, 138–185.

SUTCLIFFE, R. C. (1947). A contribution to the problem of development. *Quart. J. Roy. Meteorol. Soc.* **73**, 370–383.

SUTTON, L. J. (1925). Haboobs. *Quart. J. Roy. Meteorol. Soc.* **51**, 25–30.

SVERDRUP, H. U. (1945). Oceanography. *In* "Handbook of Meteorology" (F. A. Berry, E. Bollay, and N. R. Beers, eds.), pp. 1029–1056. McGraw-Hill, New York.

SWALLOW, J. C. (1965). The Somali Current. Some observations made aboard RRS "Discovery" during August 1964. *Mar. Observ.* **35**, 125–130.

TAKEDA, T. (1966). Effects of the prevailing wind with vertical shear on the convective cloud accompanied with heavy rain. *J. Meteorol. Soc. Japan* **44**, 129–243.

TAO, S.-Y. (1949). Movements of surface atmosphere in China (in Chinese). *Meteorol. J. Acad. Sinica* Special Issue commemorating the sixtieth birthday of Chu Co Ching.

TAYLOR, G. (1932). Climatology of Australia. *In* "Handbuch der Klimatologie" (W. Köppen and R. Geiger, eds.), Vol. 4, Part S. Borntraeger, Berlin.

THOMPSON, B. W. (1951). An essay on the general circulation of the atmosphere over Southeast Asia and the West Pacific. *Quart. J. Roy. Meteorol. Soc.* **77**, 569–597.

THOMPSON, B. W. (1957a). Some reflections on equatorial and tropical forecasting. *East African Meteorol. Dept. Tech. Mem. No.* 7.

THOMPSON, B. W. (1957b). The diurnal variation of precipitation in British East Africa. *East African Meteorol. Dept. Tech. Mem. No.* 8.

THOMPSON, B. W. (1965). "The Climate of Africa." Oxford Univ. Press, London and New York.

THOMPSON, B. W. (1968). Tables showing the diurnal variation of precipitation in East Africa and Seychelles. *East African Meteorol. Dept. Tech. Mem. No.* 10.

TREWARTHA, G. T. (1961). "The Earth's Problem Climates." Univ. of Wisconsin Press, Madison, Wisconsin.

TSCHIRHART, G. (1958). Les conditions aérologiques à l'avant des lignes des grains en Afrique équatoriale. *France, Météorol. Nat., Monogr.,* **11**.

TSCHIRHART, G. (1959). Les perturbations atmosphériques intéressant l'A.E.F. meridionale. *France, Météorol. Nat., Monogr.* **13**.

TSE, S. Y. W. (1966). A new method for the prediction of typhoon movement using the 700 mb chart. *Quart. J. Roy. Meteorol. Soc.* **92**, 239–253.

TU, C.-W. (1934). China rainfall and world weather. *Mem. Roy. Meteorol. Soc.* **4**, 100–117.

TU, C.-W. (1937). China weather and world oscillation with application to long-range forecasting of floods and droughts of China during the summer. *Acad. Sinica Nat. Res. Inst. Meteorol. Mem.* **11**, No. 4.

U.S. Committee for GARP (1969). Plan for U.S. Participation in the Global Atmospheric Research Program. Nat. Acad. of Sci., Washington, D.C.

U.S. Weather Bureau (1955–1959). U.S. Navy Marine Climatic Atlas of the World, Volumes I–V. Chief of Naval Operations, Washington, D.C.

VALOVCIN, F. R. (1970). An objective method of forecasting changes in convective activity in Southeast Asia. *In Proc. Conf. Summer Monsoon Southeast Asia* (C. S. Ramage, ed.), pp. 215–222. Navy Weather Res. Facility, Norfolk, Virginia.

VAN BEMMELEN, W. (1905). Staubnebel im Malayischen Archipel im Jahre 1902. *Meteorol. Z.* **22**, 362–365.

VANDERMAN, L. W., AND COLLINS, W. G. (1967). Operational-experimental numerical forecasting for the tropics. *Mon. Weather Rev.* **95**, 950–953.

VAN DUIJNEN MONTIJN, J. A. (1952). Indian Ocean Oceanographic and Meteorological Data, 2nd ed. Roy. Netherlands Meteorol. Inst., de Bilt, No. 135.

VAN LOON, H., AND TALJAARD, J. J. (1958). A study of the 1000-500 mb thickness distribution in the Southern Hemisphere. *Notos* 7, 123–158.

VEDERMAN, J., HIRATA, G. H., AND MANNING, E. J. (1966). Forecasting in the tropics with a barotropic atmospheric model. *Mon. Weather Rev.* 94, 337–344.

VERMILLION, C. H. (1969). Weather Satellite Picture Receiving Stations. Inexpensive Construction of Automatic, Picture Transmission Ground Equipment. NASA, Washington, D.C.

VERPLOEGH, G. (1960). On the annual variation of climatic elements of the Indian Ocean. *Kon. Ned. Meteorol. Inst. Mededel. Verhand. No. 77.*

WALKER, G. T. (1910–1916). Correlations in seasonal variations of weather, III. On the criterion for the reality of relationships or periodicities. *Mem. India Meteorol. Dept.* 21, Part 9, 12–15.

WALKER, G. T. (1923). Correlations in seasonal variations of weather, VIII. A preliminary study of world weather. *Mem. India Meteorol. Dept.* 24, 75–131.

WALKER, G. T. (1924). Correlations in seasonal variations of weather, X. Applications to seasonal forecasting in India. *Mem. India Meteorol. Dept.* 24, 333–345.

WALKER, H. O. (1960). The monsoon in West Africa. *In* "Monsoons of the World." (S. Basu *et al.*, eds.), pp. 35–42. Hind Union Press, New Delhi.

WANG, R.-H. (1963). Tracks of extratropical cyclones in East Asia (in Chinese). *Acta Meteorol. Sinica* 33, 15–24.

WARD, R. D. (1925). "The Climate of the United States." Ginn, Boston, Massachusetts.

WARK, D. Q., AND HILLEARY, D. T. (1969). Atmospheric temperature: Successful testing of remote probing. *Science* 165, 1256–1258.

WARNECKE, G., AND SUNDERLIN, W. S. (1968). The first color picture of the earth taken from the ATS-3 satellite. *Bull. Amer. Meteorol. Soc.* 49, 75–83.

WATTS, I. E. M. (1955). "Equatorial weather." Oxford Univ. Press (Univ. London), London and New York.

WEICKMANN, L. Jr. (1963). Mittlere Luftdruckverteilung im Meeresniveau während der Hauptjahreszeiten im Bereiche um Afrika, in dem Indischen Ozean und den angrenzenden Teilen Asiens. *Meteorol. Runds.* 16, 89–100.

WEIGHTMAN, R. H. (1941). Preliminary studies in seasonal weather forecasting. *Mon. Weather Rev. Suppl. No. 45.*

WELDON, R. B. (1969). Meso-scale distribution of convective activity in the Saigon area as related to synoptic regimes. (to be published).

WEXLER, H. (1951). Anticyclones. *Compendium Meteorol.*, 621–629.

WIDGER, W. K., SHERR, P. E., AND ROGERS, C. W. C. (1964). Practical Interpretation of Meteorological Satellite Data. Aracon Geophys., Contr. USAF 19(628)-2471, Final Rep.

WIEDERANDERS, C. J. (1961). Analyses of Monthly Mean Resultant Winds for Standard Pressure Levels over the Pacific. Hawaii Inst. Geophys. Rep. 13.

WISSMANN, H. V. (1939). Die Klima-und Vegetationsgebiete Eurasiens. *Z. Ges. Erdk. Berlin* 1–14.

WOOSTER, W. S., AND REID, J. L. (1963). Eastern boundary currents. *In* "The Sea" (M. N. Hill, ed.), pp. 253–280. Wiley, New York.

WOOSTER, W. S., SCHAEFER, M. B., AND ROBINSON, M. K. (1967). Atlas of the Arabian Sea for Fishery Oceanography. Univ. California, Inst. Marine Resources, La Jolla, California.

World Meteorological Organization (1957). Commission for Instruments and Observation. Abridged final rep., second session, Paris. Geneva.

World Meteorological Organization (1968). Research Work in Tropical Meteorology, Eighth Report. Geneva.

WRIGHT, P. B., AND EBDON, R. A. (1968). Upper air observations at the Seychelles. 1963–64. *Geophys. Mem.* **15**, 1–85.

WYRTKI, K. (1956). The rainfall over the Indonesian waters. *Kementerian Perhubungan Lembaga Meteorologi dan Geofisik Verhand. No.* 49.

WYRTKI, K. (1957). Precipitation, evaporation and energy exchange at the surface of the Southeast Asian waters. *In* Marine Research in Indonesia No. 3, pp. 12–40. Inst. Marine Res., Djakarta, Indonesia.

WYRTKI, K. (1966). Seasonal Variation of Heat Exchange and Surface Temperature in the North Pacific Ocean. Hawaii Inst. Geophys. Rep. 66-3.

WYRTKI, K. (1969). Personal communication.

YANAI, M. (1964). Formation of tropical cyclones. *Rev. Geophys.* **2**, 367–414.

YEH, T. C. (1949). On energy dispersion in the atmosphere. *J. Meteorol.* **6**, 1–16.

YEH, T. C. (1950). The circulation of the high troposphere over China in the winter of 1945–46. *Tellus* **2**, 173–183.

YEH, T.-C., DAO, S.-Y., and LI, M.-T. (1959). The abrupt change of circulation over the Northern Hemisphere during June and October. *In* "The Atmosphere and the Sea in Motion" (B. Bolin, ed.), pp. 249–267. Rockefeller Inst., New York.

YIN, M. T. (1949). A synoptic-aerologic study of the summer monsoon over India and Burma. *J. Meteorol.* **6**, 393–400.

YOSHINO, M. M. (1963). Rainfall, frontal zones and jet streams in early summer over East Asia. *Bonner Meteorol. Abhand. No.* 3.

YOSHINO, M. M. (1966). Four stages of the rainy season in early summer over East Asia (Part II). *J. Meteorol. Soc. Japan, Ser. II* **44**, 209–217.

YOSHINO, M. M. (1967). Atmospheric circulation over the North-west Pacific in summer. *Meteorol. Runds.* **20**, 45–52.

ZAIN, Z. (1969). Personal communication.

ZIPSER, E. J., AND TAYLOR, R. C. (1968). A Catalogue of Meteorological Data Obtained During the Line Islands Experiment February–April 1967. Nat. Center Atmos. Res., Boulder and Hawaii Inst. of Geophys., Honolulu, Hawaii.

ZOU, H., QIAN, Z-Q., ZHU, C-Y., AND Qiang, P-Q. (1964). An analysis of the 500 mb circulation in the middle and lower Yangtze Valley during the Mei-Yü period (in Chinese). *Acta Meteorol. Sinica* **34**, 174–184.

Author Index

Numbers in italics refer to the pages on which the complete references are listed.

279

Subject Index

International Geophysics Series

Editor

J. VAN MIEGHEM

Royal Belgian Meteorological Institute
Uccle, Belgium